LONDON MATHEMATICAL SOCIETY LECTURE NOTE SERIES

Managing Editor: Professor I.M.James, Mathematical Institute, 24-29 St.Giles, Oxford

This series publishes the records of lectures and seminars on advanced topics in mathematics held at universities throughout the world. For the most part, these are at post-graduate level either presenting new material or describing older matter in a new way. Exceptionally, topics at the undergraduate level may be published if the treatment is sufficiently original.

Prospective authors should contact the editor in the first instance.

Already published in this series

1. General cohomology theory and K-theory, PETER HILTON.
4. Algebraic topology: A student's guide, J. F. ADAMS.
5. Commutative algebra, J. T. KNIGHT.
7. Introduction to combinatory logic, J. R. HINDLEY, B. LERCHER and J. P. SELDIN.
8. Integration and harmonic analysis on compact groups, R. E. EDWARDS.
9. Elliptic functions and elliptic curves, PATRICK DU VAL.
10. Numerical ranges II, F. F. BONSALL and J. DUNCAN.
11. New developments in topology, G. SEGAL (ed.).
12. Symposium on complex analysis Canterbury, 1973, J. CLUNIE and W. K. HAYMAN (eds.).
13. Combinatorics, Proceedings of the British combinatorial conference 1973, T. P. McDONOUGH and V. C. MAVRON (eds.).
14. Analytic theory of abelian varieties, H. P. F. SWINNERTON-DYER.
15. An introduction to topological groups, P. J. HIGGINS.
16. Topics in finite groups, TERENCE M. GAGEN.
17. Differentiable germs and catastrophes, THEODOR BRÖCKER and L. LANDER.
18. A geometric approach to homology theory, S. BUONCRISTIANO, C. P. ROURKE and B. J. SANDERSON.
19. Graph theory, coding theory and block designs, P. J. CAMERON and J. H. VAN LINT.
20. Sheaf theory, B. R. TENNISON.
21. Automatic continuity of linear operators, ALLAN M. SINCLAIR.
22. Presentations of groups, D. L. JOHNSON.
23. Parallelisms of complete designs, PETER J. CAMERON.
24. The topology of Stiefel manifolds, I. M. JAMES.
25. Lie groups and compact groups, J. F. PRICE.
26. Transformation groups: Proceedings of the conference in the University of Newcastle upon Tyne, August 1976, CZES KOSNIOWSKI.

continued overleaf

27. Skew field constructions, P. M. COHN.
28. Brownian motion, Hardy spaces and bounded mean oscillation, K. E. PETERSEN.
29. Pontryagin duality and the structure of locally compact abelian groups, SIDNEY A. MORRIS.
30. Interaction models, N. L. BIGGS.
31. Continuous crossed products and type III von Neumann algebras, A. VAN DAELE.

London Mathematical Society Lecture Note Series. 32

Uniform Algebras and Jensen Measures

T. W. GAMELIN

Professor of Mathematics
University of California
Los Angeles

CAMBRIDGE UNIVERSITY PRESS
CAMBRIDGE
LONDON NEW YORK MELBOURNE

CAMBRIDGE UNIVERSITY PRESS
Cambridge, New York, Melbourne, Madrid, Cape Town, Singapore, São Paulo

Cambridge University Press
The Edinburgh Building, Cambridge CB2 8RU, UK

Published in the United States of America by Cambridge University Press, New York

www.cambridge.org
Information on this title: www.cambridge.org/9780521222808

First published 1978
Re-issued in this digitally printed version 2008

A catalogue record for this publication is available from the British Library

ISBN 978-0-521-22280-8 paperback

Contents

Preface

These notes are based on lectures given in various courses and seminars over past years. The unifying theme is the notion of subharmonicity with respect to a uniform algebra. Dual to the generalized subharmonic functions are the Jensen measures.

Chapter 1 includes an abstract treatment of Jensen measures, which also includes the standard basic elements of Choquet theory. It is based on an approach of D.A.Edwards. Chapter 2 shows how the various classes of representing measures fit into the abstract setting, and Chapter 3 deals specifically with the algebra $R(K)$.

In Chapter 4, we present an example due to B.Cole of a Riemann surface R which fails to be dense in the maximal ideal space of $H^{\infty}(R)$.

Chapter 5 is based upon recent work of N.Sibony and the author concerning algebras generated by Hartogs series, and the abstract Dirichlet problem for function algebras. The abstract development is applied in Chapter 6 to algebras of analytic functions of several complex variables. Here the generalized subharmonic functions turn out to be closely related to the plurisubharmonic functions, and the abstract Dirichlet problem turns out to be Bremermann's generalized Dirichlet problem.

Chapters 7 and 8 are devoted to Cole's theory of the conjugation operator in the setting of uniform algebras. The problem is to determine which of the classical estimates relating a trigonometric polynomial and its conjugate extend to the abstract setting. Cole shows that many inequalities fail to extend to arbitrary representing measures, while "all" inequalities extend to the context of Jensen measures.

In Chapter 9, the problem of characterizing the moduli of the functions in $H^2(\sigma)$ is considered. The discussion is based on Cole's proof of a theorem of Helson, which frees Helson's theorem from the underlying group structure.

References are given at the end of each chapter. At the very end of the notes, there is an index of symbols.

In preparing these notes, I have benefited from mathematical contacts with a number of people. Let me acknowledge first and foremost my debt to Brian Cole. His incisive ideas and remarkable results form the basis for a sizeable portion of these lecture notes. Special thanks go to Don Marshall, for writing up one of the preliminary versions of Chapter 9.

I would like to thank Julie Honig for her excellent work typing the penultimate version of the manuscript. And I would like to thank the staff at the Cambridge University Press for facilitating the publication of these notes.

T.W.Gamelin

Harcourt Hill
Oxford
1978

1 · Choquet theory

Here the basic ideas of Choquet theory are developed in a framework suitable for uniform algebras. The lectures of R. Phelps[6] provide a very readable account of Choquet theory, as does also the expository paper of G.Choquet and P.A. Meyer[3]. Their approach has been modified by D.A.Edwards[4], in order to handle Jensen measures and the Jensen-Hartogs inequality for function algebras. We will follow the development of Choquet and Meyer, as amended by Edwards.

$\mathcal{R}$-measures

Let M be a compact space, and let $\mathcal{R}$ be a family of continuous functions from M to the extended line $[-\infty,+\infty)$. We will assume always that $\mathcal{R}$ has the following properties.

$$\mathcal{R} \text{ includes the constant functions.} \tag{1.1}$$

$$\text{If } m \in Z_+ \text{ and } v,w \in \mathcal{R}\text{, then } (v+w)/m \in \mathcal{R}. \tag{1.2}$$

$$\mathcal{R} \text{ separates the points of } M. \tag{1.3}$$

An *$\mathcal{R}$-measure* for $\phi \in M$ is a probability measure σ on M such that

$$w(\phi) \le \int w d\sigma, \quad w \in \mathcal{R}. \tag{1.4}$$

Since $\mathcal{R}$ includes the constants, the estimate (1.4) is equivalent to the estimate

$$\int w d\sigma \ge 0, \text{ for all } w \in \mathcal{R} \text{ such that } w(\phi) = 0. \tag{1.5}$$

As an example, observe that the point mass δ_ϕ at ϕ is always an $\mathcal{R}$-measure for ϕ .

The theory applies to any linear subset $\mathcal{R}$ of $C_R(M)$ that contains the constants and separates the points of M . In this case, the fact that $\mathcal{R} = -\mathcal{R}$ implies that the $\mathcal{R}$-measures for a point $\phi \in M$ are the representing measures for ϕ , that is, the probability measures σ on M that satisfy

$$u(\phi) = \int u d\sigma , \quad \text{all} \quad u \in \mathcal{R} .$$

In the principal application dealt with by Choquet theory, M is a compact convex subset of a locally convex linear topological vector space, and $\mathcal{R}$ is the space of continuous real-valued affine functions on M . In this case, each probability measure σ on M is an $\mathcal{R}$-measure for some $\phi \in M$, ϕ being referred to as the "barycenter" of σ .

The main example that will occupy our attention is the case in which $\mathcal{R}$ consists of functions of the form $(\log |f|)/m$, where m is a positive integer, and f belongs to an algebra A of continuous complex-valued functions on some compact space M . In this case, a probability measure σ on M is an $\mathcal{R}$-measure for $\phi \in M$ if and only if the Jensen-Hartogs inequality is valid:

$$\log |f(\phi)| \le \int \log |f| \, d\sigma , \quad f \in A .$$

The $\mathcal{R}$-measures are called *Jensen measures*.

Now fix $\phi \in M$, and fix a compact subset X of M . Let $\mathcal{U}$ be the set of functions $u \in C_R(X)$ such that $u > w$ for some $w \in \mathcal{R}$ satisfying $w(\phi) = 0$. If the functions in $\mathcal{R}$ are continuous, then $\mathcal{U}$ is simply the algebraic sum of the positive continuous functions on X , and the functions in $\mathcal{R}$ vanishing at ϕ .

On account of (1.2), $\mathcal{U}$ is a convex cone. Since $0 \in \mathcal{R}$, the cone $\mathcal{U}$ includes the positive functions in $C_R(X)$.

Since the restriction of every $w \in \mathcal{R}$ to X is a lower envelope of functions in $\mathcal{U}$, a probability measure σ on X is an $\mathcal{R}$-measure if and only if $\int u d\sigma \geq 0$ for all $u \in \mathcal{U}$. In particular, the $\mathcal{R}$-measures can be described by inequalities involving integrals of continuous functions, so that the set of $\mathcal{R}$-measures on X is a convex, weak-star compact set.

1.1 Lemma. Let X be a compact subset of M, and let $\phi \in M$. There exists an $\mathcal{R}$-measure σ for ϕ supported on X if and only if

$$w(\phi) \leq \sup_{x \in X} w(x), \quad w \in \mathcal{R}. \tag{1.6}$$

Proof. If there is an $\mathcal{R}$-measure σ for ϕ on X, then the inequality $w(\phi) \leq \int_X w d\sigma$ yields (1.6) immediately. Conversely, if (1.6) is valid, then the constant function -1 does not belong to $\mathcal{U}$, and $\mathcal{U}$ is a proper cone in $C_R(X)$. By the separation theorem for convex sets, there is a nonzero measure τ on X such that $\int u d\tau \geq 0$ for all $u \in \mathcal{U}$. Since $\mathcal{U}$ includes the positive functions, τ is a positive measure. The measure $\sigma = \tau/\tau(X)$ is then a probability measure that is nonnegative on $\mathcal{U}$, so that σ is an $\mathcal{R}$-measure. □

The following version of the monotone extension theorem is due in this setting to D.A.Edwards[4].

1.2 Theorem (Edwards' Theorem). Let $\phi \in M$, and let X be a compact subset of M. For each lower semi-continuous function Q from X to $(-\infty, +\infty]$, the following quantities are equal:

$$\sup\{w(\phi) : w \in \mathcal{R},\ w \le Q \text{ on } X\}, \tag{1.7}$$

$$\inf\{\int Q d\sigma : \sigma \text{ is an } \mathcal{R}\text{-measure for } \phi \text{ on } X\}. \tag{1.8}$$

Here the infimum in (1.8) is declared to be $+\infty$ if there is no $\mathcal{R}$-measure for ϕ on X.

Proof. Let S denote the supremum in (1.7) and let I denote the infimum in (1.8). If there are no $\mathcal{R}$-measures for ϕ on X, there is by Lemma 1.1 a function $w \in \mathcal{R}$ such that $w(\phi) > 0$, while $w < 0$ on X. Multiplying w by a large positive constant, we can arrange that $w \le Q$ on X, while $w(\phi)$ is arbitrarily large. Hence $S = +\infty$, so that $S = I$.

We can assume then that there exist $\mathcal{R}$-measures for ϕ on X. If σ is such an $\mathcal{R}$-measure, and if $w \in \mathcal{R}$ satisfies $w \le Q$, then $w(\phi) \le \int w d\sigma \le \int Q d\sigma$. Hence $S \le I$.

To prove the reverse inequality, suppose first that Q is continuous. Let $b > S$. Then there is no $w \in \mathcal{R}$ such that $w(\phi) = 0$ and $w + b \le Q$. Consequently $Q - b$ does not belong to the cone $\mathcal{U}$ defined earlier. By the separation theorem for convex sets, there is a nonzero measure τ on X such that τ is nonnegative on $\mathcal{U}$, while $\int (Q-b) d\tau \le 0$. Since $\tau \ge 0$ on $\mathcal{U}$, $\sigma = \tau/\tau(X)$ is an $\mathcal{R}$-measure. Furthermore, $\int (Q-b) d\sigma \le 0$, so that $\int Q d\sigma \le b$, and $I \le b$. Since $b > S$ is arbitrary, we conclude that $I = S$, in the case at hand.

For the general case, consider a continuous function $q \le Q$, and let $I_q = S_q$ denote the quantity above determined by q. For each such q, choose an $\mathcal{R}$-measure σ_q such that $\int q d\sigma_q = S_q$. This choice is possible; since the set of $\mathcal{R}$-measures is weak-star compact. Let σ be a weak-star adherent point of the net $\{\sigma_q\}_{q<Q}$ as q increases to Q. If q is fixed, and the continuous function p satisfies $q \le p \le Q$, then $\int q d\sigma_p \le \int p d\sigma_p = S_p \le S$. Letting p

increase to Q , we obtain $\int q d\sigma \le S$. Since $q \le Q$ is arbitrary, $\int Q d\sigma \le S$. Hence $I = S$, and moreover there exists an R-measure σ such that $\int Q d\sigma = I$. □

Applying Edwards' Theorem to the function $-\chi_E$, where χ_E is the characteristic function of a closed subset E of X , we obtain the following corollary.

1.3 Corollary. Suppose there is an R-measure on X for ϕ . Then for any closed subset E of X ,

$$\sup\{\sigma(E) : \sigma \text{ an } R\text{-measure on } X \text{ for } \phi\}$$

is equal to

$$\inf\{u(\phi) : u \in -R,\ u \ge 0 \text{ on } X,\ u \ge 1 \text{ on } E\} .$$

The Family S of R-envelope Functions

A lower semi-continuous function u from a compact subset X of M to $(-\infty,+\infty]$ is an R-*envelope function* on X if u is the upper envelope on X of functions in R . The family of R-envelope functions on M is denoted by S . In the abstract framework, the R-envelope functions often play the role that the subharmonic functions play in classical potential theory.

Every R-envelope function on X is evidently the restriction to X of a function in S .

If $c \in \mathbb{R}$, and if $w_1,\ldots,w_n \in R$, then the function

$$\max(c,w_1,\ldots,w_n) \tag{1.9}$$

is a continuous R-envelope function. The functions in S are simply the upper envelopes of functions of the form (1.9). An elementary compactness argument establishes the following.

1.4 Lemma. The continuous R-envelope functions are the uniform limits of functions of the form $\max(c, w_1, \ldots, w_n)$, where c is real and $w_1, \ldots, w_n \in R$.

From the definitions, we see immediately that S has the following properties.

$$S \text{ includes the constant functions.} \tag{1.10}$$

$$\text{If } u \in S \text{ and } c > 0 \text{, then } cu \in S . \tag{1.11}$$

$$\text{If } u, v \in S \text{, then } u + v \in S . \tag{1.12}$$

$$\text{If } \{u_\alpha\} \text{ is any subset of } S \text{, then } \sup_\alpha u_\alpha \text{ belongs to } S . \tag{1.13}$$

The following property of S is sufficiently important to merit a separate statement.

1.5 Lemma. If $u \in S$, and if χ is an increasing continuous convex function from an interval containing the range of u to $(-\infty, +\infty]$, then $\chi \circ u \in S$.

Proof. There is only one possible point of discontinuity of an arbitrary increasing convex function χ, namely, a point β such that $\chi(t) = +\infty$ for $t > \beta$, while $\chi(t) < +\infty$ for $t < \beta$. We must assume that $\chi(t)$ tends to $\chi(\beta)$ as t increases to β. In this case, χ is an upper envelope of functions of the form $at + b$, where $a > 0$. Consequently $\chi \circ u$ is an upper envelope of functions of the form $au + b$, where $a > 0$. Since each of these belongs to S, so does $\chi \circ u$. □

The R-envelope functions are dual, in some sense, to R-measures. This duality is exhibited by the following characterization of the R-envelope functions.

1.6 Theorem. Let u be a lower semi-continuous function from M to $(-\infty,+\infty]$. Then u is an R-envelope function if and only if

$$u(\phi) \le \int u d\sigma \tag{1.14}$$

for all $\phi \in M$ and all R-measures σ on M for ϕ.

Proof. Since (1.14) holds for all $u \in R$, it also holds for all upper envelopes of functions in R, hence for all $u \in S$.

Conversely, suppose that (1.14) is valid for all $\phi \in M$ and all R-measures σ on M for ϕ. Let v be any continuous function on M such that $v < u$. According to Edwards' Theorem (Theorem 1.2), there exists for each $\phi \in M$ and each $\varepsilon > 0$, a function $w \in R$ such that $w < u$, while $w(\phi) > u(\phi)-\varepsilon$. It follows that u is an upper envelope of functions in R. □

From Theorem 1.6 and Fatou's Lemma, we obtain immediately the following.

1.7 Corollary. If $\{u_j\}_{j=1}^{\infty}$ is a sequence in S that is bounded above, and if $u = \limsup_{j\to\infty} u_j$ is bounded and lower semi-continuous, then $u \in S$.

There is another simple proof of Lemma 1.5, based on Theorem 1.6 and Jensen's inequality. Recall that *Jensen's inequality* is the estimate

$$\chi\left(\int u d\sigma\right) \le \int \chi \circ u \, d\sigma \,, \tag{1.15}$$

valid whenever σ is a probability measure, u is real-valued and χ is an increasing convex real-valued function of a real variable. The validity of (1.15) for simple functions boils down to the convexity of χ.

To prove Lemma 1.5, one notes that if σ is an R-measure for ϕ, then $\chi(u(\phi)) \le \chi(\int u d\sigma) \le \int \chi \circ u \, d\sigma$, so that by Theorem 1.6, $\chi \circ u \in S$.

The Family S_C of Continuous R-envelope Functions

We denote by S_C the family of (finite) R-envelope functions on M that are continuous. As observed earlier in Lemma 1.4, these are the uniform limits of the functions of the form (1.9).

Evidently S_C is a convex cone that separates points and contains the constants. Consequently S_C enjoys the properties (1.1), (1.2) and (1.3) postulated for R. The theory we have developed can be applied to S_C in place of R. Observe though that the S_C-measures are precisely the R-measures, while the S_C-envelope functions coincide with the R-envelope functions. For many purposes, the family R can be replaced by the family S_C.

An important property enjoyed by S_C, but not necessarily by R, is that of being a semi-lattice. The maximum of any two functions in S_C again belongs to S_C. This leads to the following observation, which plays a crucial role in the treatment of maximal measures.

1.8 Lemma. The algebraic difference $S_C - S_C$ is dense in $C_R(M)$.

Proof. If $v_1, v_2, w_1, w_2 \in S_C$, then

$$\max(v_1 - w_1, v_2 - w_2) = \max(v_1 + w_2, v_2 + w_1) - w_1 - w_2 .$$

It follows that $S_C - S_C$ is a lattice. Since it separates points and contains the constants, it is dense in $C_R(M)$, by the lattice version of the Stone-Weierstrass Theorem. □

The $\mathcal{R}$-Dirichlet Problem

Let u be a lower semi-continuous function from M to $(-\infty,+\infty]$. In analogy to the procedure followed by Perron to solve the Dirichlet problem, we define the (lower) *solution to the $\mathcal{R}$-Dirichlet problem* with data u on M to be the upper envelope $\tilde{u}$ of the functions in $\mathcal{R}$ dominated by u on M :

$$\tilde{u}(\phi) = \sup\{w(\phi) : w \in \mathcal{R}, w < u \text{ on } M\} . \tag{1.16}$$

The supremum defining $\tilde{u}$ could as well be taken over the functions in $\mathcal{S}$, or in $\mathcal{S}_C$, dominated by u . Edwards' Theorem gives an alternative expression for $\tilde{u}$:

$$\tilde{u}(\phi) = \inf\{\int u d\sigma : \sigma \text{ an } \mathcal{R}\text{-measure on } M \text{ for } \phi\} .$$

Some elementary properties of the correspondence $u \to \tilde{u}$ are as follows:

$$\tilde{u} \text{ is an } \mathcal{R}\text{-envelope function ,} \tag{1.17}$$

$$\tilde{u} = u \text{ if and only if } u \in \mathcal{S} , \tag{1.18}$$

$$\widetilde{cu} = c\tilde{u} \text{ if } c > 0 , \tag{1.19}$$

$$\tilde{u} + \tilde{v} \le \widetilde{u+v} , \tag{1.20}$$

$$\tilde{u} \le \tilde{v} \text{ whenever } u \le v , \tag{1.21}$$

if a net $\{u_\alpha\}$ of lower semi-continuous functions increases pointwise on M to u , then $\tilde{u}_\alpha$ increases pointwise on M to $\tilde{u}$. (1.22)

In applications, we will wish to consider a lower semi-continuous boundary function u defined only on a compact subset X of M. Again $\tilde{u}$ is defined to be the upper envelope on M of the functions in S dominated by u on X. This amounts to declaring u to be $+\infty$ on $M\backslash X$, and defined $\tilde{u}$ as before. From Edwards' Theorem, we obtain

$$\tilde{u}(\phi) = \inf\{\int u d\sigma : \sigma \text{ an } R\text{-measure on } X \text{ for } \phi\}, \quad (1.23)$$

where the infimum is declared to be $+\infty$ if there are no R-measures for ϕ on X.

The Choquet Boundary

The *Choquet boundary* of R, denoted by ∂_R, consists of those points $\phi \in M$ such that the point mass δ_ϕ at ϕ is the only R-measure for ϕ.

By Edwards' Theorem, any continuous real-valued function u on M satisfies $\tilde{u} = u$ on the Choquet boundary. This property characterizes the Choquet boundary. Indeed, suppose ϕ is not a Choquet boundary point, and choose an R-measure σ for ϕ such that $\sigma \neq \delta_\phi$. Then any $u \in C_R(M)$ satisfying $\int u d\sigma < u(\phi)$, also satisfies $\tilde{u}(\phi) < u(\phi)$.

The next lemma shows that the notion of Choquet boundary point is independent, in the appropriate sense, of the compact set on which the functions in R are defined.

1.9 Lemma. Let X be a closed subset of M such that every $w \in R$ attains its maximum on X, and let $x_0 \in X$. If the point mass at x_0 is the only R-measure for x_0 on X, then x_0 belongs to the Choquet boundary of R.

Proof. Let σ be an R-measure for x_0 on M. Let E be a compact neighbourhood of x_0 in X, and let u be a continuous function on X such that $u \leq 0$, $u(x_0) = 0$, and

$u \leq -1$ on $X \backslash E$. Consider the $\mathcal{R}$-envelope function $\tilde{u}$ on M. From (1.23), we obtain $\tilde{u} \leq 0$ on M, while $\tilde{u}(x_0) = 0$. The estimate $\tilde{u}(x_0) \leq \int u d\sigma$ then shows that $\tilde{u} = 0$ a.e. $(d\sigma)$. Suppose ϕ belongs to $\text{supp}(\sigma)$, the closed support of σ, and $\tilde{u}(\phi) = 0$. Let μ be an $\mathcal{R}$-measure on X for ϕ. The estimates $0 = \tilde{u}(\phi) \leq \int \tilde{u} d\mu \leq \int u d\mu$ show that $u = 0$ on $\text{supp}(\mu)$, so that μ is supported on E. Since such ϕ are dense in $\text{supp}(\sigma)$, and since any weak-star limit of $\mathcal{R}$-measures is an $\mathcal{R}$-measure, we see that every $\phi \in \text{supp}(\sigma)$ has an $\mathcal{R}$-measure supported on E. Since E is an arbitrary compact neighbourhood of x_0, we conclude that the point mass at x_0 is an $\mathcal{R}$-measure for every point of $\text{supp}(\sigma)$. Hence $w(\phi) \leq w(x_0)$ for all $\phi \in \text{supp}(\sigma)$ and all $w \in \mathcal{R}$. Since $w(x_0) \leq \int w d\sigma$ for all $w \in \mathcal{R}$, each $w \in \mathcal{R}$ assumes the constant value $w(x_0)$ on $\text{supp}(\sigma)$. Since $\mathcal{R}$ separates points, σ is the point mass at x_0. □

E. Bishop and K. de Leeuw[2] have given an example, a porcupine space, for which the Choquet boundary is not a Borel set. This adverse behaviour does not occur when M is metrizable.

1.10 Lemma. If M is metrizable, then the Choquet boundary $\partial_{\mathcal{R}}$ is a G_δ-set.

Proof. Let $\{u_j\}_{j=1}^{\infty}$ be a dense sequence in $C_{\mathbb{R}}(X)$. Evidently a point $\phi \in M$ belongs to the Choquet boundary if and only if $\tilde{u}_j(\phi) = u_j(\phi)$ for all j. Since each set $\{\tilde{u}_j = u_j\}$ is the intersection of the open sets $\{u_j - \tilde{u}_j < 1/n\}$, $n \geq 1$, we see that $\partial_{\mathcal{R}}$ is itself a countable intersection of open sets. □

1.11 Theorem. Each $u \in S_C$ attains its maximum at a Chouquet boundary point, as does each $w \in \mathcal{R}$.

Proof. The latter assertion follows by applying the former to a function of the form $u = \max(c,w)$, where c is very negative.

The proof of the first assertion is modelled on a standard proof of the Krein-Milman Theorem. We begin by introducing an auxiliary notion, corresponding in convexity theory to a face of a convex set.

A closed subset E of M is an *R-face* if every R-measure for each point of E is supported by E . Evidently M is itself an R-face. A singleton $\{\phi\}$ is an R-face if and only if ϕ is a Choquet boundary point.

Let $u \in S_C$, and let α be the maximum value of u on M . Suppose $\phi \in M$ satisfies $u(\phi) = \alpha$. If σ is an R-measure for ϕ , then the estimates $u \le \alpha$ and $\alpha = u(\phi) \le \int u d\sigma$ show that $u = \alpha$ on the support of σ . Hence the set on which any $u \in S_C$ attains its maximum is an R-face.

The R-faces evidently form an inductive family, when ordered by inclusion. By Zorn's Lemma, every R-face includes a minimal R-face. Therefore it suffices to show that every minimal R-face consists of one point.

Suppose that F is a minimal R-face. Let $v \in S_C$, let F_0 be the subset of F on which v attains its maximum, let $\phi \in F_0$, and let σ be an R-measure on M for ϕ . Since F is an R-face, σ is supported on F . Furthermore, σ is an $R|_F$-measure for ϕ , where $R|_F$ is the restriction of R to F . Since v is an $R|_F$-envelope function, the subset F_0 of F on which v attains its maximum is an $R|_F$-face. Hence σ is supported on F_0 . It follows that F_0 is an R-face. Since F is minimal, F_0 coincides with F , and v is constant on F . Since $v \in S_C$ is arbitrary, and since S_C separates the point of M , F reduces to a single point. □

As a corollary, we obtain a version of Shilov's Theorem on

the existence of minimal closed boundaries.

1.12 Theorem. There exists a smallest closed subset of R, namely the closure of the Choquet boundary of R, on which each function in R attains its maximum.

Proof. If X is any closed subset of R on which each function in R attains its maximum, then by Lemma 1.1 all $\phi \in M$ have an R-measure on X. Therefore X includes ∂_R, hence the closure of ∂_R. On the other hand, Theorem 1.11 shows that every $w \in R$ attains its maximum on ∂_R. □

We will refer to the closure of ∂_R as the *Shilov boundary* associated with R.

Barriers

In analogy with classical potential theory, we say that an R-envelope function u on M is a *barrier* at $\phi \in M$ if $u \le 0$, $u(\phi) = 0$ and $u < 0$ on $M \backslash \{\phi\}$. If there is a barrier u at ϕ, then ϕ is a Choquet boundary point. Indeed, the estimates $u \le 0$ and $0 = u(\phi) \le \int u d\sigma$ show that any R-measure σ for ϕ is concentrated at ϕ.

There is also a topological condition that must be satisfied for ϕ to have a barrier. If u is a barrier at ϕ, then the intersection of the open sets $\{u > -1/n\}$, $n \ge 1$, includes only ϕ, so that the singleton $\{\phi\}$ is a G_δ-set.

Conversely, if ϕ is a Choquet boundary point which forms a G_δ-set, then there is a barrier at ϕ. Indeed, if v is any continuous function on M such that $v \le 0$, $v(\phi) = 0$ and $v < 0$ on $M \backslash \{\phi\}$, then $\tilde{v}$ is a barrier at ϕ.

In the setting of classical potential theory, M.V. Keldysh[5] has proved that any point having a barrier actually has a continuous barrier. This theorem extends to our general setting. In fact we have the following characterization of Choquet boundary points.

1.13 Theorem. The following are equivalent, for a point $\phi_0 \in M$.

(i) ϕ_0 belongs to the Choquet boundary of R .

(ii) If h is any continuous real-valued function on M , then there exists a continuous R-envelope function u on M such that $u \le h$ while $u(\phi_0) = h(\phi_0)$.

(iii) There exist $\alpha < 0 < \beta$ with the following property: for any compact subset E of M not containing ϕ_0 , there exists $v \in R$ such that $v(\phi_0) = 0$, $v \le \alpha$ on E , and $v \le \beta$ on M .

Proof. Evidently (ii) implies (iii), for any $\alpha < 0 < \beta$.

Suppose that (iii) is valid, and suppose furthermore that (i) fails. Let σ be an R-measure for ϕ_0 distinct from the point mass δ_0 at ϕ_0 . Write $\sigma = a\tau + (1-a)\delta_0$, where τ is a probability measure with no mass at ϕ_0 , and $0 < a \le 1$. Let E be a compact set not including ϕ_0 , and let v be the corresponding function from (iii). Then $0 = v(\phi_0) \le \int v d\sigma = a\int v d\tau \le a\alpha\tau(E) + a\beta\tau(M\backslash E)$. If E is chosen, though, so that most of the mass of τ is carried by E , this last sum is negative, and we obtain a contradiction. It follows that (iii) implies (i).

For the remaining implication, assume that (i) is valid, and let $h \in C_R(M)$. Fix real numbers b and c such that

$$b < \min h \le \max h < c .$$

We claim that for any compact subset E of M not including ϕ_0 , there exists $v \in R$ such that $v(\phi_0) = h(\phi_0)$, $v \le c$, and $v \le b$ on E . Indeed, let $\varepsilon > 0$ be small, and let $g \in C_R(M)$ satisfy $g(\phi_0) = h(\phi_0)$, $g \le h(\phi_0)$ on M , and $g \le b - \varepsilon$ on E. Since $\tilde{g}(\phi_0) = h(\phi_0)$, there exists $w \in R$ such that $w \le g$, while $w(\phi_0) = h(\phi_0) - \varepsilon$. If ε is sufficiently small, then $v = w + \varepsilon$ has the desired properties.

Now fix $0 < s < 1$, and choose a sequence $\{\varepsilon_m\}_{m=1}^{\infty}$ which decreases rapidly to zero. Precise conditions on the ε_m's and on s will be specified momentarily.

We construct by induction a sequence $\{u_j\}_{j=0}^{\infty}$ in R as follows. Let u_0 be the constant function $h(\phi_0)$. Suppose $u_0, \ldots, u_{m-1}$ have been chosen so that $u_j(\phi_0) = h(\phi_0)$, $0 \le j \le m-1$. The compact set

$$E_m = \{\phi : \max_{0 \le j \le m-1} u_j(\phi) \ge h(\phi) + \varepsilon_m\}$$

then does not contain ϕ_0. By our preliminary observation, we may choose $u_m \in R$ so that $u_m(\phi_0) = h(\phi_0)$, $u_m \le c$, and $u_m \le b$ on E_m.

Now consider the series

$$u = (1 - s) \sum_{j=0}^{\infty} s^j u_j .$$

If the u_j's are not uniformly bounded below, we replace u_j by $\max(-\gamma, u_j) \in S_C$, when γ is a large constant. Then the u_j's are uniformly bounded, and this insures that the series converges uniformly on M to a function $u \in S_C$. Furthermore, $u(\phi_0) = h(\phi_0)$. We must show that $u \le h$.

If $u_j(\phi) \le h(\phi)$ for all j, then certainly $u(\phi) \le h(\phi)$. Suppose that $u_j(\phi) > h(\phi)$ for some index j. Then $\phi \in \cup_{k=1}^{\infty} E_k$, and since the E_k's are increasing, there is a first index $m \ge 0$ such that $\phi \in E_{m+1}$ while $\phi \notin E_m$. Then $u_j(\phi) < h(\phi) + \varepsilon_m$ for $0 \le j \le m-1$, while $u_m(\phi) \le c$, and $u_j(\phi) \le b$ for $j > m$. Substituting these estimates in the series defining u, and taking into account the appropriate modifications in the case $m = 0$, we obtain

$$u(\phi) \le (1-s)\left[(h(\phi)+\varepsilon_m)\sum_{j=0}^{m-1} s^j + s^m c + b\sum_{j=m+1}^{\infty} s^j\right]$$

$$= (1-s^m)(h(\phi)+\varepsilon_m) + (1-s)s^m c + s^{m+1} b$$

$$= h(\phi) + (1-s^m)\varepsilon_m + s^m[(1-s)c + sb - h(\phi)] .$$

If we choose $s < 1$ so near 1 that

$$(1-s)c + sb - \min h < 0 ,$$

and then we choose the ε_m's so small that

$$(1-s^m)\varepsilon_m + s^m[(1-s)c + sb - \min h] < 0 , \quad m \ge 1 ,$$

we obtain the estimate $u(\phi) < h(\phi)$. □

The idea used in the preceding proof stems from Bishop[1,2].

Maximal Measures

We define a partial ordering on the probability measures on M , by declaring $\mu \prec \nu$ to mean that $\int u d\mu \le \int u d\nu$ for all $u \in S_C$. This is equivalent to requiring that $\int w d\mu \le \int w d\nu$ for all $w \in R$. The relation "$\prec$" is evidently symmetric and transitive. Since by Lemma 1.8 the linear space generated by S_C is dense in $C_R(M)$, the relations $\mu \prec \nu$ and $\nu \prec \mu$ together imply that $\mu = \nu$.

As an example, observe that the probability measures σ on M such that satisfy $\delta_\phi \prec \sigma$ are precisely the R-measures for the point $\phi \in M$. Heuristically speaking, the larger a probability measure is with respect to this ordering, the more it is dispersed towards the boundary.

A probability measure on M is *maximal* if it is maximal

with respect to the ordering "<". Since the conditions defining the ordering are weak-star continuous, a weak-star compactness argument shows that the set of probability measures, with the ordering "<", is inductive. By Zorn's Lemma, every probability measure is dominated by a maximal probability measure. In particular, every point $\phi \in M$ has a maximal $\mathcal{R}$-measure.

Our aim now is to show that maximal measures are supported "near" the Choquet boundary. We begin by stating a version of the Hahn-Banach Theorem that is convenient for our purposes.

1.14 Theorem. Let p be a sublinear functional on a real vector space V, that is, p is a real-valued function on V such that

$$p(u+v) \le p(u) + p(v), \qquad u,v \in V$$

$$p(cu) = cp(u), \qquad u \in V, c \ge 0.$$

Then there exists a linear functional L on V such that $L \le p$. Furthermore, L is unique if and only if p is linear, in which case $L = p$.

Proof. We consider only the uniqueness assertion.

Let V_0 be a one-dimensional subspace of V, generated by a nonzero vector v. A linear function L_0 on V_0 is dominated by p if and only if $L_0(v) \le p(v)$ and $L_0(-v) \le p(-v)$, that is, if and only if $-p(-v) \le L_0(v) \le p(v)$. Defining first L_0 on V_0, and then using the Hahn-Banach Theorem to extend L_0 to V we see that any real value in the interval $[-p(-v),p(v)]$ is assumed at v by some linear functional L on V dominated by p. The dominated extension is then unique if and only if $p(v) = -p(-v)$ for all $v \in V$, and in this case $L = p$. □

1.15 Lemma. Let ν be a probability measure on M, and define a sublinear functional p by

$$p(u) = -\int (\widetilde{-u})d\nu , \quad u \in C_R(M) .$$

The functionals L on $C_R(M)$ dominated by p are precisely those arising from probability measures μ satisfying $\nu \prec \mu$.

Proof. Suppose first that $L \le p$. Since $-p(-1) = p(1) = 1$, also $L(1) = 1$. If $u \le 0$, then $-u \ge 0$, $(\widetilde{-u}) \ge 0$, $p(u) \le 0$, and $L(u) \le 0$. It follows that $L \ge 0$, so that L is represented by a probability measure μ. If $u \in S_C$, then $\int u d\nu = \int \tilde{u} d\nu = -p(-u) \le -L(-u) = L(u) = \int u d\mu$. Hence $\nu \prec \mu$.

Conversely, suppose that $\nu \prec \mu$. Let $v,w \in S_C$. It is easy to verify the identity $\widetilde{w-v} = w + (\widetilde{-v})$, and from this we obtain $-p(v-w) = \int (\widetilde{w-v})d\nu \le \int (\widetilde{w-v})d\mu = \int w d\mu + \int (\widetilde{-v})d\mu \le \int (w-v)d\mu = -L(v-w)$. According to Lemma 1.8, such functions $v-w$ are dense in $C_R(M)$. Hence $-p(u) \le -L(u)$ for all $u \in C_R(M)$, and $L \le p$. □

Combining the two preceding lemmas, we arrive easily at the main result of this section.

1.16 Theorem. A probability measure ν on M is maximal if and only if for each $u \in C_R(M)$, ν is carried by the G_δ-set $\{u = \tilde{u}\}$.

Proof. If ν is maximal, then by Lemma 1.15, ν is the unique linear functional on $C_R(X)$ dominated by the sublinear function p defined in Lemma 1.15. By the uniqueness clause of the Hahn-Banach Theorem, $\nu = p$. Hence $\int u d\nu = -p(-u) = \int \tilde{u} d\nu$ for all $u \in C_R(X)$. Since $\tilde{u} \le u$, ν is carried on the set $\{u = \tilde{u}\}$.

Conversely, if ν is carried by each set $\{u = \tilde{u}\}$, then $\int u d\nu = \int \tilde{u} d\nu$ for all $u \in C_R(X)$. Hence $p(v) = \int v d\nu$ for all $v \in C_R(X)$, and ν is the unique linear functional on $C_R(X)$ dominated by p. By Lemma 1.15, ν is maximal.

□

As a simple application of Theorem 1.16, we observe that every maximal measure ν is supported by the Shilov boundary. Indeed, suppose that $p \in M$ does not belong to the Shilov boundary. Choose $u \in C_R(M)$ such that $u(p) > 0$, while $u = 0$ on the Shilov boundary. Then $\tilde{u} \leq 0$. Theorem 1.16 implies that the closed support of ν is contained in the set $\{u \leq 0\}$, and hence does not include p.

Now each of the sets $\{\tilde{u} = u\}$ is a G_δ-set that includes $\partial_{\mathcal{R}}$, and $\partial_{\mathcal{R}}$ is the intersection of these sets. The question arises as to whether the maximal measures are precisely those probability measures carried by $\partial_{\mathcal{R}}$. This fleeting hope fails spectacularly. The examples of Bishop and de Leeuw cited earlier show that $\partial_{\mathcal{R}}$ need not be a Borel set. Moreover, even if $\partial_{\mathcal{R}}$ is a Borel set, there may exist maximal $\mathcal{R}$-measures with no mass on $\partial_{\mathcal{R}}$.

In the positive direction, Bishop and de Leeuw have shown that a maximal measure has zero mass on any Baire set that is disjoint from $\partial_{\mathcal{R}}$. From this, they obtain easily their generalization of Choquet's Theorem to the nonmetrizable case. In our setting, their result asserts that for each $\phi \in M$, there is a probability measure μ on the σ-ring generated by $\partial_{\mathcal{R}}$ and the Baire sets, such that $\mu(\partial_{\mathcal{R}}) = 1$, and $u(\phi) \leq \int u d\mu$ for all $u \in \mathcal{S}$.

In the case that M is metrizable, the proof of Lemma 1.10 shows that $\partial_{\mathcal{R}}$ is the intersection of a sequence of sets of the form $\{\tilde{u} = u\}$. From Theorem 1.16 we deduce that the maximal measures are the probability measures carried by $\partial_{\mathcal{R}}$. In particular, we have the following version of Choquet's Theorem.

1.17 Theorem. Suppose that M is metrizable. Then every $\phi \in M$ has an R-measure that is carried by the Choquet boundary ∂_R.

Examples

For the most trivial example, we take M to be an interval $[\alpha,\beta]$ on the real line, and we take R to be the set of functions of the form $c + at$, where c is real and $a \geq 0$. In this case, the R-envelope functions are the continuous convex increasing functions from $[\alpha,\beta]$ to $(-\infty,+\infty]$. The Choquet boundary consists of the singleton $\{\beta\}$.

More generally, let M be any compact subset of $\mathbb{R}^n$, and let R be the set of functions of the form $u(t) = c + \Sigma_{j=1}^n a_j t_j$ where c is real and $a_j \geq 0$, $1 \leq j \leq n$. The functions in S are convex. If we provide $\mathbb{R}^n$ with the ordering for which $s \leq t$ means $s_j \leq t_j$, $1 \leq j \leq n$, then the functions in S are increasing. The Choquet boundary consists of those extreme points of the closed convex hull of E which are maximal with respect to this ordering of $\mathbb{R}^n$.

Next let us return to the principal example for convexity theory, in which M is a compact convex subset of a locally convex real linear topological vector space V, and R is the family of continuous affine functions on M. These are the uniform limits on M of functions of the form $c + L$, where c is a real constant and L is a continuous linear functional on V. Recall that a probability measure σ on M is an R-measure for ϕ if and only if σ represents ϕ, that is, ϕ is the barycenter of σ. The functions in S_C turn out to be the continuous convex real-valued functions on M, and the Choquet boundary of R coincides with the set of extreme points of M. Theorem 1.17 specializes to Choquet's Theorem: If M is metrizable, then the set of extreme points of M is a G_δ-set, and every $\phi \in M$ is the barycenter of a probability measure on the set of extreme points of M.

References

1. Bishop, E. A minimal boundary for function algebras, *Pac. J. Math.* 9 (1959), 629-642.
2. Bishop, E. and de Leeuw, K. The representation of linear functionals by measures on sets of extreme points, *Ann. Inst. Fourier (Grenoble)* 9 (1959), 305-331.
3. Choquet, G. and Meyer, P.A. Existence et unicité des representations intégrals dans les convexes compacts quelconques, *Ann. Inst. Fourier (Grenoble)* 13 (1963), 139-154.
4. Edwards, D.A. Choquet boundary theory for certain spaces of lower semicontinuous functions, in *Function Algebras*, F. Birtel (ed.), Scott, Foresman and Co., 1966, pp.300-309.
5. Keldysh, M.V. On the solubility and stability of Dirichlet's problem, *Uspekhi Mat. Nauk USSR* 8 (1941), 171-231.
6. Phelps, R.R. *Lectures on Choquet's Theorem, Van Nostrand Mathematical Studies* No.7, Van Nostrand, 1966.

2 · Classes of representing measures

Throughout these notes, we will denote by A a uniform algebra on a compact space X. In other words, A is a closed subalgebra of $C(X)$ containing the constants and separating the points of X. The maximal ideal space of A will be denoted by M_A, and A will be regarded as an algebra of functions on M_A.

Associated with A, there are three natural choices for the space $\mathcal{R}$ of the preceding chapter. We will consider in turn these three choices. The $\mathcal{R}$-measures become respectively the representing measures, the Arens-Singer measures, and the Jensen measures.

Representing Measures

For this example, we take $\mathcal{R}$ to be the space $\mathrm{Re}(A)$ of real parts of functions in A. Since $\mathcal{R}$ is a linear space, a probability measure σ on M_A is an $\mathcal{R}$-measure for $\phi \in M_A$ if and only if $u(\phi) = \int u d\sigma$ for all $u \in \mathrm{Re}(A)$. This is equivalent to

$$f(\phi) = \int f d\sigma , \quad f \in A . \tag{2.1}$$

A probability measure σ on M_A that satisfies (2.1) is said to be a *representing measure for* ϕ. The Choquet boundary of $\mathrm{Re}(A)$ is the set of points for which the point mass is the only representing measure.

The Edwards Theorem (Theorem 1.2) specializes immediately to the following useful theorem.

2.1 Theorem. If $\phi \in M_A$, and if Q is a lower semi-continuous function from X to $(-\infty,+\infty]$ then

$$\sup\{u(\phi) : u \in \text{Re}(A),\ u < Q \text{ on } X\}, \tag{2.2}$$

is equal to

$$\inf\{\int Q d\sigma : \sigma \text{ a representing measure for } \phi \text{ on } X\}. \tag{2.3}$$

As an application of Theorem 1.11, we prove the existence of a "Shilov boundary" for A.

2.2 Theorem. There is a smallest closed subset E of M_A such that every function in A attains its maximum modulus on E.

Proof. Let E be the closure of the Choquet boundary of $\text{Re}(A)$. By Theorem 1.11, every $u \in \text{Re}(A)$ attains its maximum modulus on E, and consequently every $f \in A$ attains its maximum modulus on E.

Let F be another closed subset of M_A such that every $f \in A$ attains its maximum modulus on F. Fix $\phi \in M_A$. Then

$$|f(\phi)| \leq \sup_{x \in F} |f(x)|, \quad f \in A.$$

Hence there is a measure σ on F such that $\|\sigma\| = 1$, while (2.1) holds. Applying (2.1) to the function 1, we find that σ is a probability measure, so that σ is a representing measure for ϕ on F. Since $u(\phi) = \int u d\sigma$ for $u \in \text{Re}(A)$, each $u \in \text{Re}(A)$ attains its maximum on F. By Theorem 1.11, F includes the Choquet boundary of $\text{Re}(A)$, and $F \supseteq E$. □

The subset E of Theorem 2.2 is called the *Shilov boundary of* A , and it is denoted by ∂_A . Evidently $\partial_A \subseteq X$.

Theorems 2.1 and 2.2 depend only on the linear structure of A . They are valid for any closed separating subspace of C(X) containing the constants. Now we turn to a characterization of the Choquet boundary points of Re(A) as the generalized peak points of A . This result depends on the algebraic structure of A .

We begin by defining several concepts related to peak sets and peak points. For more background and details, see [4].

A closed subset E of M_A is a *peak set* if there is a function $f \in A$ such that $f = 1$ on E , while $|f| < 1$ on $M_A \backslash E$. The function f is said to *peak on* E . A closed set E is a *generalized peak set* if it is an intersection of peak sets. It is easy to show that a generalized peak set is a peak set if and only if it is a G_δ-set .

A point $\phi \in M_A$ is a *peak point* for A if the singleton $\{\phi\}$ is a peak set, and it is a *generalized peak point* if $\{\phi\}$ is a generalized peak set.

2.3 Theorem. The following are equivalent, for a uniform algebra A on M_A , and $\phi_0 \in M_A$.

(i) ϕ_0 is a generalized peak point.

(ii) ϕ_0 belongs to the Choquet boundary of Re(A) .

(iii) There exist $0 < a < 1$ and $c \geq 1$ with the following property. For each compact subset E of M_A not including ϕ_0 , there is $g \in A$ satisfying $g(\phi_0) = 1$, $\|g\| \leq c$, and $|g| \leq a$ on E .

(iv) For each strictly positive $h \in C_R(M_A)$, there exists $f \in A$ such that $f(\phi_0) = h(\phi_0)$ while $|f| \leq h$ on M_A .

Proof. Suppose that (iii) is valid. If g is as in (iii), then the function $u = Re(g) - 1$ in Re(A) satisfies

$u(\phi_0) = 0$, $u \le c$, and $u \le a-1$ on E . Consequently the condition (iii) of Theorem 1.13 is met, and ϕ_0 is a Choquet boundary point for $\mathrm{Re}(A)$. Of course, it is easy to show directly from (iii) that the point mass at ϕ_0 is the only representing measure for ϕ_0 . In any event, (iii) implies (ii).

By taking powers of peaking functions, one sees that (i) implies (iii), with $c = 1$ and a arbitrarily small. Since (iv) obviously implies (i), it remains to prove that (ii) implies (iv). We will mimic the proof of the corresponding implication of Theorem 1.13.

Suppose then that (ii) is valid. Choose b and c such that

$$0 < b < \min h \le \max h < c .$$

We claim that for any compact subset E of M_A not including ϕ_0 , there exists $f \in A$ such that $f(\phi_0) = h(\phi_0)$, $|f| \le c$, and $|f| \le b$ on E . Indeed, using Theorem 2.1 as in the proof of Theorem 1.13, we find $v = \mathrm{Re}(g)$ in $\mathrm{Re}(A)$ such that $v(\phi_0) = \log h(\phi_0)$, $v < \log c$, and $v < \log b$ on E . If g is normalized so that $g(\phi_0)$ is real, then $f = e^g$ has the desired properties.

Now we proceed as in the proof of Theorem 1.13, fixing $0 < s < 1$ and a sequence $\{\varepsilon_m\}_{m=1}^{\infty}$ decreasing rapidly to zero. We construct by induction a sequence $\{f_j\}_{j=1}^{\infty}$ in A as follows. Let $f_0 = h(\phi_0)$ be constant. Having chosen $f_0,\dots,f_{m-1}$ so that $f_j(\phi_0) = h(\phi_0)$, $0 \le j \le m-1$, set

$$E_m = \{\phi : \max_{0\le j\le m-1} |f_j(\phi)| \ge h(\phi) + \varepsilon_m\} ,$$

and choose $f_m \in A$ so that $f_m(\phi_0) = h(\phi_0)$, $|f_m| \le c$, and $|f_m| \le b$ on E_m . The series

$$f = (1-s) \sum_{j=0}^{\infty} s^j f_j$$

converges uniformly, so that $f \in A$, and $f(\phi_0) = h(\phi_0)$. The estimates in the proof of Theorem 1.13, with u_j replaced by $|f_j|$, show that $|f| \le h$ providing s and $\{\varepsilon_m\}$ are chosen properly. □

Arens-Singer Measures

As the second application, we take R to be the set of functions

$$\left\{\frac{1}{m} \log |f| : m \in Z_+,\ f \in A^{-1}\right\}, \tag{2.4}$$

where A^{-1} denotes the group of invertible elements of A. Evidently R includes the constant functions, and R separates the points of M_A. From the identities

$$\log |f^{-1}| = -\log |f|,$$

$$\frac{1}{m_1} \log |f_1| + \frac{1}{m_2} \log |f_2| = \frac{1}{m_1 m_2} \log \left|f_1^{m_2} f_2^{m_1}\right|,$$

we see that R is a linear space over the rational numbers. In particular, R satisfies (1.1) through (1.3), and the theory developed in Chapter 1 applies to R.

Since $R = -R$, any R-measure σ on M_A for ϕ satisfies $u(\phi) = \int u d\sigma$ for $u \in R$. Consequently the R-measures for ϕ are the probability measures σ on M_A such that

$$\int \log |f| d\sigma = \log |f(\phi)|, \quad f \in A^{-1}. \tag{2.5}$$

Such measures are called *Arens-Singer measures*.

There is another way to view Arens-Singer measures. It is easy to check that the functional

$$L(\log |g|) = \log |g(\phi)| , \quad g \in A^{-1} ,$$

extends to a continuous linear functional on the closed linear span of $\log |A^{-1}|$. The Arens-Singer measures correspond to the positive extensions of this functional from the linear span of $\log |A^{-1}|$ to $C_R(M_A)$.

By applying (2.5) to e^g , where $g \in A$ is arbitrary, we find that

$$\int \mathrm{Re}(g)d\sigma = \mathrm{Re}(g(\phi)) , \quad g \in A ,$$

so that every Arens-Singer measure for ϕ is a representing measure for ϕ . This can also be deduced by observing that the family (2.4) defining the Arens-Singer measures for ϕ includes the family $\mathrm{Re}(A)$ defining the representing measures for ϕ .

The *Arens-Singer boundary points* are defined to be the points for which the point mass is the only Arens-Singer measure. These points are now characterized by Theorem 1.13.

What is the Shilov boundary associated with this choice of R ? Since R includes $\mathrm{Re}(A)$, every generalized peak point for A is an Arens-Singer boundary point. On the other hand, every function in R evidently assumes its maximum on the Shilov boundary ∂_A of A . It follows that the Shilov boundary associated with R coincides with ∂_A . In particular, every $\phi \in M_A$ has an Arens-Singer measure supported on ∂_A .

There is one class of examples for which every representing measure is an Arens-Singer measure. If M_A is simply connected, in the sense that the Čech cohomology group $\check{H}^1(M_A;Z) = \{0\}$, then according to the Arens-Royden Theorem, every invertible function in A is an exponential. In this case, the family R of (2.4), defining the Arens-Singer measures, coincides with $\mathrm{Re}(A)$.

The third choice for R , leading to the most important application for our purposes, is the family of functions

$$\{\frac{1}{m} \log |f| : f \in A, m \in Z_+\} . \tag{2.6}$$

These functions are continuous, from M_A to the extended line $[-\infty,+\infty)$. Again (1.1), (1.2) and (1.3) are easy to establish, so that the theory developed in Chapter 1 applies. It is convenient to note that we could as well work with the cone of functions of the form $c \log |f|$, where $c > 0$ and $f \in A$.

In this case, a probability measure σ on M_A is an R-measure for ϕ if and only if

$$\log |f(\phi)| \le \int \log |f| d\sigma , \quad f \in A . \tag{2.7}$$

Such measures are called *Jensen measures*, and the inequality (2.7) will be referred to as the *Jensen-Hartogs inequality*.

Every Jensen measure is an Arens-Singer measure. This follows from the observation that the family (2.6) leading to the Jensen measures includes the family (2.4) leading to the Arens-Singer measures. One can also obtain (2.5) directly by applying the Jensen-Hartogs inequality to f and $1/f$.

The Choquet boundary associated with the family (2.6) is called the *Jensen boundary* of A . These are the points for which the point mass is the only Jensen measure. The Jensen boundary includes the Arens-Singer boundary. Since every function in the family (2.6) assumes its maximum on ∂_A , the Shilov boundary associated with the family (2.6) coincides with the Shilov boundary ∂_A of A .

The R-envelope functions associated with the family (2.6) will be referred to as the *log-envelope functions*. Thus a function w from a compact subset E of M_A to $(-\infty,+\infty]$ is a *log-envelope function on* E if w is bounded below, and w is an upper envelope of functions of the form $c \log |f|$, where $c > 0$ and $f \in A$. The Jensen-Hartogs inequality remains valid for log-envelope functions.

The R-Dirichlet problem associated with this family of functions will be referred to as the *A-Dirichlet problem.* Thus the solution to the A-Dirichlet problem with boundary data u on the compact subset E of M_A is the upper envelope $\tilde{u}$ of all functions of the form $c \log |f|$, where $c > 0$ and $f \in A$ satisfy $c \log |f| < u$ on E. We will return to study log-envelope functions and the A-Dirichlet problem in some detail in Chapter 5.

By Lemma 1.5, the composition of an increasing convex function χ with any log-envelope function is again a log-envelope function. In particular,

$$\chi(\log |f(\phi)|) \leq \int \chi(\log |f(x)|)d\sigma(x) , \quad f \in A , \tag{2.8}$$

for any Jensen measure σ for ϕ. This estimate follows directly from the Jensen-Hartogs inequality and Jensen's inequality. In the special case $\chi(t) = e^{pt}$, where $p > 0$ is fixed, (2.8) leads to

$$|f(\phi)| \leq \left[\int |f(x)|^p d\sigma(x)\right]^{1/p} , \quad f \in A , \ p > 0 . \tag{2.9}$$

Hence the evaluation functional at ϕ is continuous in the L^p-metric, for all $p > 0$.

The estimate (2.9) actually characterizes Jensen measures among the probability measures on X. Indeed, the limit relation

$$\lim_{p\to 0+} \left[\int |f(x)|^p d\sigma(x)\right]^{1/p} = e^{\int \log|f(x)|d\sigma(x)}$$

shows that (2.9) becomes the Jensen-Hartogs inequality (2.7) as p decreases to zero.

For future reference, we state the specialization of Theorem 1.13 to the case at hand.

2.4 Theorem. The following are equivalent, for a point $x_0 \in X$.

(i) x_0 is a Jensen boundary point.

(ii) If h is any continuous real-valued function on X, then there exists a continuous log-envelope function u on X such that $u \le h$, while $u(x_0) = h(x_0)$.

(iii) There exist $\alpha < 0 < \beta$ with the following property: for any compact subset E of X not containing x_0, there exist $f \in A$ and $c > 0$ such that $f(x_0) = 1$, $c \log|f| \le \alpha$ on E, and $c \log|f| \le \beta$ on X.

An Example

For the simplest nontrivial example, we take X to be the boundary $\partial\Delta$ of the open unit disc Δ, and A to be the disc algebra $A(\Delta)$ consisting of the analytic functions on Δ that extend continuously to $\partial\Delta$. The maximal ideal space of $A(\Delta)$ coincides with the closed unit disc $\bar{\Delta}$. In this case, $\mathrm{Re}(A)$ is dense in $C_R(\partial\Delta)$. Hence each point in $\bar{\Delta}$ has a unique representing measure on $\partial\Delta$, and this representing measure is necessarily a Jensen measure. For the representing measure $\frac{d\theta}{2\pi}$ for the origin, the inequality (2.7) becomes the usual inequality reflecting the subharmonicity of $\log|f|$,

$$\log|f(0)| \le \int \log|f(e^{i\theta})|\frac{d\theta}{2\pi}, \qquad f \in A(\Delta). \tag{2.10}$$

In turn, (2.10) serves to generate inequalities of the form

$$\chi(\log|f(0)|) \le \int \chi(\log|f(e^{i\theta})|)\frac{d\theta}{2\pi}, \qquad f \in A(\Delta), \tag{2.11}$$

valid whenever χ is an increasing convex function.

There is a converse assertion of sorts, that shows that the Jensen-Hartogs inequality (2.10) is essentially the best inequality that can be expected. It asserts that if χ is a real-valued Borel function of a real variable such that (2.11) is valid, then χ is an increasing convex function. Let us prove this assertion.

Suppose that (2.11) is valid. If $a < b$, there exists $f \in A(\Delta)$ such that $|f| = e^b$ on $\partial\Delta$, while $f(0) = e^a$. Substituting this f in (2.11), we obtain $\chi(a) \le \chi(b)$, so that χ is an increasing function. To establish the convexity of χ, first observe that the inequality (2.11) is readily established for all $f \in H^\infty(\Delta)$. Let $s,t \in \mathbb{R}$, and let $0 < c < 1$. Let u be a real-valued function on $\partial\Delta$ that assumes the value s on a set of mass c, and the value t on the complementary set of mass $1 - c$. Extend u harmonically to Δ, let $*u$ be the conjugate harmonic function of u, and set $f = \exp(u+i*u)$. Then

$$\log |f(0)| = \int u \frac{d\theta}{2\pi} = cs + (1-c)t ,$$

and (2.11) becomes

$$\chi(cs + (1-c)t) \le c\chi(s) + (1-c)\chi(t) .$$

Hence χ is convex, and our assertion is established.

Nomenclature

We close with some remarks on the history of Jensen measures and the Jensen-Hartogs inequality (2.7).

In 1899, J.L.W.V.Jensen published a paper in the Acta Mathematica "Sur un nouvel et important théorème de la théorie des fonctions". The important theorem that Jensen refers to is now known as the Jensen formula, that

$$\frac{1}{2\pi}\int_0^{2\pi} \log|f(re^{i\theta})|d\theta = \log|f(0)| + \log\left[\frac{r^{n-m}|\beta_1|\ldots|\beta_m|}{|\alpha_1|\ldots|\alpha_n|}\right],$$

where f is meromorphic on the closed disc of radius r, with zeros $\alpha_1,\ldots,\alpha_n$ and poles $\beta_1,\ldots,\beta_m$. Jensen applied his formula to entire functions, in order to relate the growth rate and the distribution of zeros. From the Jensen formula one readily obtains the inequality (2.10) in the case that f is analytic. However, the estimate (2.10) does not appear explicitly in the paper.

Apparently Jensen was not the first to write down the Jensen formula. J.Hadamard[5, p.50] asserts that he had discovered the Jensen formula already in 1888. He did not publish the result, though, as he was waiting to be able to derive some significant consequences from the formula.

In their fundamental paper in this area, R.Arens and I.M.Singer[1] refer to (2.10) as Szegö's inequality, citing the paper [10] in which G.Szegö establishes the integrability of $\log|f|$ for $f \in H^2(d\theta)$. Again while (2.10) is immediately obtainable from Szegö's work, it does not appear explicitly in the paper.

More recently, K.Hoffman[7], and also V.S.Vladimirov[11], refer to (2.7) and (2.10) as Jensen's inequality. The nomenclature is justifiable, but it leads to confusion since the name has already been attached to the inequality (1.15), proved by Jensen[9] in 1906.

S.Bochner and W.T.Martin[3] refer to the estimate (2.10) as the Jensen-Hartogs inequality, in view of the seminal research of F.Hartogs[6] which hinged upon the subharmonicity of $\log|f|$. We have adopted the terminology of Bochner and Martin.

The term "Jensen measure" is due to Bishop[2], who established the existence of Jensen measures in general by roughly the same argument we have used in Lemma 1.1.

References

1. Arens, R. and Singer, I.M. Function values as boundary integrals, *P.A.M.S.* 5 (1954), 735-745.
2. Bishop, E. Holomorphic completions, analytic continuations, and the interpolation of seminorms, *Ann. of Math.* 78 (1963), 468-500.
3. Bochner, S. and Martin, W.T. *Several Complex Variables*, Princeton University Press, Princeton, 1948.
4. Gamelin, T.W. *Uniform Algebras*, Prentice-Hall, 1969.
5. Hadamard, J. *The Psychology of Invention in the Mathematical Field*, Princeton University Press, 1949.
6. Hartogs, F. Zur Theorie der analytischen Funktionen mehrer unabhängiger Veränderlichen, insbesondere über die Darstellung derselben durch Reihen, welche nach Potenzen einer Veränderlichen fortschreiten, *Math. Ann.* 62 (1906), 1-88.
7. Hoffman, K. Analytic functions and logmodular Banach algebras, *Acta Math.* 108 (1962), 271-317.
8. Jensen, J.L.W.V. Sur un nouvel et important théorème de la théorie des fonctions, *Acta Math.* 22 (1899), 359-364.
9. Jensen, J.L.W.V. Sur les fonctions convexes et les inégalités entre les valeurs moyennes, *Acta Math.* 30 (1906), 175-193.
10. Szegö, G. Über die Randwerte einer analytischen Funktion, *Math. Ann.* 84 (1921), 232-244.
11. Vladimirov, V.S. *Methods of the Theory of Functions of Several Complex Variables*, M.I.T. Press, Cambridge, Mass., 1966.

3 · The algebra R(K)

Let K be a compact subset of the complex plane $\mathbb{C}$, and let $R(K)$ denote the uniform closure in $C(K)$ of the rational functions with poles off K. Then $R(K)$ is a uniform algebra on K, and the maximal ideal space of $R(K)$ coincides with K. We wish to specialize the discussion of the various classes of representing measures introduced earlier to the algebra $R(K)$. Pertinent background information on $R(K)$ is contained in [2].

Recall that the Cauchy transform $\hat{\nu}$ of a measure ν supported on K is defined by

$$\hat{\nu}(\zeta) = \int \frac{d\nu(z)}{z - \zeta}, \qquad \zeta \in \mathbb{C},$$

providing the integral converges. Since $\hat{\nu}$ is the convolution of ν and the locally integrable $(dxdy)$ function $1/z$, the integral defining $\hat{\nu}$ converges absolutely on a set of full Lebesgue measure. Furthermore, $\hat{\nu}$ is analytic off the closed support of ν, and $\hat{\nu}(\infty) = 0$.

It is easy to approximate a rational function with poles off K by linear combinations of functions of the form $z \to 1/(z-\zeta)$, $\zeta \in \mathbb{C}\backslash K$. Consequently the linear span of the functions $1/(z-\zeta)$, $\zeta \in \mathbb{C}\backslash K$, is dense in $R(K)$. Dual to this statement is the fact that a measure ν on K is orthogonal to $R(K)$ if and only if $\hat{\nu} = 0$ on $\mathbb{C}\backslash K$. This points to the important role played by the Cauchy transform in rational approximation theory.

Related to the Cauchy transform of a measure ν on K is the logarithmic potential of ν. We will work with the negative V_ν of the logarithmic potential, defined by

$$V_\nu(\zeta) = \int \log |z-\zeta| d\nu(z) , \quad \zeta \in \mathbb{C} . \tag{3.1}$$

The kernel $\log |z|$ appearing in the convolution integral is locally integrable $(dxdy)$, so that the integral converges absolutely on a set of full Lebesgue measure. The function V_ν is harmonic off the closed support of ν, and

$$V_\nu(\zeta) = \nu(K) \log |\zeta| + o(1) \tag{3.2}$$

as ζ tends to ∞. If ν is positive, then V_ν is (upper semicontinuous and) subharmonic.

The potential V_ν is related to the Cauchy transform of ν by the formula

$$\frac{\partial}{\partial\zeta} V_\nu = -\frac{1}{2}\hat{\nu} , \tag{3.3}$$

which is to be interpreted in the sense of distributions. We will only use (3.3) on $\mathbb{C}\backslash K$, where the formula is easy to establish by differentiating by hand. Indeed, locally one can express

$$\log |\zeta-z| = \frac{1}{2}[\log(\zeta-z) + \log(\bar{\zeta}-\bar{z})] ,$$

from which one obtains

$$\frac{\partial}{\partial\zeta} \log |\zeta-z| = \frac{1}{2}\frac{1}{\zeta-z} , \quad \zeta \neq z . \tag{3.4}$$

Differentiating the expression (3.1) for V_ν, and substituting (3.4), one obtains (3.3).

3.1 Theorem. Let $p \in K$, and let ν be a probability measure on K. Then the following are equivalent:

(i) ν is a representing measure for p,

(ii) $\hat{\nu}(\zeta) = 1/(\zeta-p)$, $\zeta \in \mathbb{C}\backslash K$,

(iii) $V_\nu(\zeta) - \log|\zeta-p|$ is constant on each component of $\mathbb{C}\backslash K$.

Proof. From (3.3) and (3.4) we have

$$\frac{\partial}{\partial\zeta}[V_\nu(\zeta) - \log|\zeta-p|] = \frac{1}{2}\left[\frac{1}{\zeta-p} - \hat{\nu}(\zeta)\right] .$$

Since $V_\nu(\zeta) - \log|\zeta-p|$ is real, the function is constant just as soon as its ζ-derivative vanishes. Hence (ii) and (iii) are equivalent.

The identity (ii) means precisely that ν represents p on the functions $z \to 1/(\zeta-z)$, $\zeta \in \mathbb{C}\backslash K$. As observed earlier, the linear span of these functions is dense in $R(K)$, so that (i) and (ii) are equivalent. □

It is easy to establish directly the equivalence of (i) and (iii) in Theorem 3.1, without passing through the Cauchy transform. Suppose for instance that ν represents p, and that $\zeta_0 \in \mathbb{C}\backslash K$. For each fixed ζ near ζ_0, there is a single-valued determination of $\log[(z-\zeta)/(z-\zeta_0)]$ that belongs to $R(K)$. For this determination we have

$$\int \log\left(\frac{z-\zeta}{z-\zeta_0}\right)d\nu(z) = \log\left(\frac{p-\zeta}{p-\zeta_0}\right) .$$

Taking real parts, we obtain

$$V_\nu(\zeta) - V_\nu(\zeta_0) = \log|p-\zeta| - \log|p-\zeta_0| .$$

This shows that $V_\nu(\zeta) - \log|p-\zeta|$ is constant on each component of $K\backslash\mathbb{C}$.

Observe that the constant value of $V_\nu(\zeta) - \log|\zeta-p|$ on the unbounded component of $\mathbb{C}\backslash K$ is zero. This follows from (3.2), which shows that

$$\lim_{\zeta\to 0} V_\nu(\zeta) - \log|\zeta-p| = 0$$

whenever ν has unit mass.

Now we turn to the Arens-Singer measures on K for a point $p \in K$. Recall that these are defined to be the probability measures on K that represent the functional "evaluation at p" on the linear span of $\log|R(K)^{-1}|$, that is, the linear span of the logarithms of the moduli of the invertible elements of $R(K)$.

Denote by $H(K)$ the closure in $C_R(K)$ of the family of functions harmonic in a neighbourhood of K.

3.2 Lemma. The closed linear span of $\log|R(K)^{-1}|$ coincides with $H(K)$.

Proof. Any function in $\log|R(K)^{-1}|$ can be approximated uniformly on K by a function of the form $\log|f|$, where f is a rational function with neither poles nor zeros on K. In particular, $\log|f|$ belongs to $H(K)$, so that the closed linear span of $\log|R(K)^{-1}|$ is included in $H(K)$.

Suppose u is harmonic in a neighbourhood of K. Choose points $z_1,\ldots,z_m \in \mathbb{C}\backslash K$ and constants $c_1,\ldots,c_m$ such that

$$w = u - \sum c_j \log|z-z_j|$$

has a single-valued harmonic conjugate $*w$ in a neighbourhood of K. Then $f = \exp(w + i{*w})$ belongs to $R(K)^{-1}$, and so do the functions $z-z_j$, $1 \le j \le m$. Hence $u = \log|f| + \Sigma c_j \log|z-z_j|$ belongs to the linear span of $\log|R(K)^{-1}|$. □

3.3 Theorem. Let $p \in K$, and let ν be a probability measure on K. The following are equivalent:

(i) ν is an Arens-Singer measure for p;

(ii) ν represents p on $H(K)$, that is $u(p) = \int u d\nu$ for every function u harmonic in a neighbourhood of K ;

(iii) $V_\nu(\zeta) = \log|\zeta - p|$ for all $\zeta \in \mathbb{C}\backslash K$.

Proof. The equivalence of (i) and (ii) follows immediately from the preceding lemma. That (ii) implies (iii) is immediate from the definition of V_ν .

Suppose that (iii) is valid. Let u be an infinitely differentiable function on $\mathbb{C}$ with compact support, such that u is harmonic in a neighbourhood of K . Then u can be represented as the logarithmic potential of its Laplacian,

$$u(\zeta) = -\frac{1}{2\pi}\int (\Delta u)(z) \log|z-\zeta| dxdy \ .$$

Since u is harmonic near K , this integral can be taken over $\mathbb{C}\backslash K$, and we obtain

$$\begin{aligned}\int u(\zeta)d\nu(\zeta) &= -\frac{1}{2\pi}\int_{\mathbb{C}\backslash K} (\Delta u)(z)\left[\int \log|z-\zeta|d\nu(\zeta)\right] dxdy \\ &= -\frac{1}{2\pi}\int (\Delta u)(z) \log|z-p| \, dxdy \\ &= u(p) \ .\end{aligned}$$

This establishes (ii), and the proof is complete. □

Finally we turn to Jensen measures. Let ν be a Jensen measure on K for p . Applying the Jensen-Hartogs inequality (2.7) to the functions $z \to z-\zeta$, we obtain

$$\log|\zeta-p| \le \int \log|\zeta-z|d\nu(z) \ , \quad \zeta \in \mathbb{C} \ . \tag{3.5}$$

Since ν is an Arens-Singer measure, the identity (iii) of Theorem 3.3 becomes

$$\log|\zeta-p| = \int \log|\zeta-z|d\nu(z) , \quad \zeta \in \mathbb{C}\backslash K . \tag{3.6}$$

It turns out that these properties characterize Jensen measures.

3.4 Theorem. Let $p \in K$, and let ν be a probability measure on K . Then the following are equivalent:

(i) ν is a Jensen measure for p ;

(ii) ν satisfies (3.5) and (3.6);

(iii) $u(p) \le \int u d\nu$ whenever u is a real-valued function that is subharmonic in a neighbourhood of K .

Proof. We have already observed that (i) implies (ii).

Suppose that (ii) is valid, and let u be subharmonic in a neighbourhood of K . By the F.Riesz Theorem, u can be expressed in the form

$$u(z) = v(z) + \int \log|z-\zeta|d\tau(\zeta) , \quad z \in K ,$$

where τ is a positive measure supported on a compact neighbourhood of K , and v is harmonic in a neighbourhood of K . Using Fubini's Theorem, we obtain

$$\begin{aligned}\int u(z)d\nu(z) &= \int v(z)d\nu(z) + \iint \log|z-\zeta|d\nu(z)d\tau(\zeta) \\ &\ge v(p) + \int \log|\zeta-p|d\tau(\zeta) \\ &= u(p) .\end{aligned}$$

Hence (ii) implies (iii).

Finally, suppose that (iii) is valid. Let f be a rational function with poles off K . Then $\log|f|$ is subharmonic in a neighbourhood of K , so that

$$\log|f(p)| \leq \int \log|f| d\nu .$$

Since such f are dense in $R(K)$, we obtain the inequality for all $f \in R(K)$, and ν is a Jensen measure for p.

□

The equivalence of (i) and (iii) in Theorem 3.4 is dual to the following statement.

3.5 Theorem. The continuous log-envelope functions on K with respect to $R(K)$ are the uniform limits on K of the functions that are continuous and subharmonic in a neighbourhood of K.

Proof. Any continuous log-envelope function u on K is a uniform limit of functions of the form $\max(c_1 \log|f_1|, \ldots, c_m \log|f_m|)$, where $c_1, \ldots, c_m > 0$ and $f_1, \ldots, f_m$ are rational functions with poles off K. In particular, u is a uniform limit of functions that are continuous and subharmonic in a neighbourhood of K.

Conversely, if u is a uniform limit of such functions, then from (iii) of Theorem 3.4 we obtain $u(p) \leq \int u d\nu$ for all $p \in K$ and all Jensen measures ν for p. By Theorem 1.6, u is a log-envelope function. □

Now consider the $R(K)$-Dirichlet problem with boundary data $u \in C_R(\partial K)$. In view of Theorem 3.5, the solution of the problem is the upper envelope $\tilde{u}$ of all continuous subharmonic functions v in a neighbourhood of K such that $v \leq u$ on ∂K. The classical proof of Perron shows that $\tilde{u}$ is harmonic on K^o. Under certain regularity conditions, $\tilde{u}$ is continuous on K and coincides with u on ∂K, so that $\tilde{u}$ is the solution to the classical Dirichlet problem. This accounts for the terminology.

The characterization of Theorems 3.3 and 3.4 yield the following theorem of A.Debiard and B.Gaveau[1].

3.6 Theorem. Let $p \in K$, and let ν be a probability measure on ∂K. Then ν is a Jensen measure for p if and only if ν is an Arens-Singer measure for p.

Proof. Suppose that ν is an Arens-Singer measure for p that is carried by ∂K. Then

$$\log|z-p| = V_\nu(z), \quad z \in \mathbb{C}\backslash K.$$

From the upper semi-continuity of V_ν, we obtain

$$\log|z-p| \le V_\nu(z), \quad z \in \partial K. \tag{3.7}$$

To show that ν is a Jensen measure, it suffices to show that this estimate holds also on K^o.

We will appeal to the following maximum principle for the logarithmic potential u of a positive measure ν supported on a closed set E (cf. [5, pp.53-54]):

$$\limsup_{E \ni z\to\zeta} u(z) = \limsup_{z\to\zeta} u(z),$$

valid for any point $\zeta \in E$ that is not an isolated point of E.

If $p \in K^o$, we easily conclude the proof. Applying the maximum principle above to $u = -V_\nu$, we obtain from (3.7) the estimate

$$\log|\zeta-p| \le \liminf_{z\to\zeta} V_\nu(z), \quad \zeta \in \partial K.$$

Since $\log|z-p|$ is subharmonic, while V_ν is harmonic on

K^o , we obtain the estimate (3.7) for $z \in K^o$, as required.

The proof is slightly more complicated if p is an arbitrary point of K . Assume for convenience that K is contained in a small disc, so that $\log|z-\zeta| < 0$ whenever z and ζ belong to some neighbourhood of K . Define

$$u_\varepsilon(z) = \int_{|\zeta-p|\geq\varepsilon} \log|z-\zeta|d\nu(\zeta) \ .$$

Then $u_\varepsilon \geq V_\nu$ near K , so that from (3.7) we obtain

$$\log|z-p| \leq u_\varepsilon(z) \ , \qquad z \in \partial K \ .$$

Furthermore, u_ε is subharmonic, and u_ε is harmonic on K^o . Applying the maximum principle again, we obtain

$$\log|z-p| \leq \liminf_{\zeta\to z} u_\varepsilon(\zeta)$$

for all $z \in \partial K$, $z \neq p$. Also $\log|z-p| < u_\varepsilon(z)$ in a neighbourhood of p . We conclude from the usual maximum principle for harmonic functions that $\log|z-p| \leq u_\varepsilon(z)$ on K^o . Letting ε tend to zero, we obtain $\log|z-p| \leq V_\nu(z)$ on K^o , as required. □

3.7 Corollary. The Jensen boundary for $R(K)$ coincides with the Arens-Singer boundary for $R(K)$, and this in turn coincides with the Choquet boundary for $H(K)$.

Proof. The first statement follows from Theorem 3.6, Lemma 1.9, and the fact that ∂K is a closed boundary for $R(K)$. The second statement follows immediately from Theorem 3.3.

□

The requirement in Theorem 3.6 that the measure ν be situated on ∂K is essential, even in the case that K is the closed unit disc $\bar{\Delta}$. Indeed, it is easy to construct a representing measure σ for 0 with positive mass at (say) $z = 1/2$. Since $\mathbb{C}\backslash\bar{\Delta}$ is connected, any representing measure for $R(\bar{\Delta})$ is an Arens-Singer measure. However, the Jensen-Hartogs inequality shows that a Jensen measure for 0 cannot have positive mass at any point other than 0 , so that σ is not a Jensen measure.

The following striking observation is due also to Debiard and Gaveau.

3.8 Theorem. Suppose that $p \in K$ does not belong to the Jensen boundary of $R(K)$. Then for each integer $m \geq 0$, the differentiation functional $f \to f^{(m)}(p)$, defined on those $f \in R(K)$ that extend to be analytic near p , extends to a continuous linear functional on $R(K)$.

Proof. For $k \geq 1$, let E_k be the intersection of $\mathbb{C}\backslash K$ and the annulus $\{2^{-m-1} \leq |z-p| \leq 2^{-m}\}$, let $\mathrm{cap}(E_k)$ be its logarithmic capacity, and let $\gamma(E_k)$ be its analytic capacity. According to the Wiener criterion (cf. [3]), the series

$$\sum_{k=1}^{\infty} k/\log[1/\mathrm{cap}(E_k)]$$

converges. In particular, the terms of the series converge to 0 . Since $\gamma(E_k) \leq \mathrm{cap}(E_k)$, we obtain

$$\lim_{k\to 0} \frac{k}{\log[1/\gamma(E_k)]} = 0 .$$

Let M be large. Then $k/\log[1/\gamma(E_k)] \leq 1/M$ for k large, so that

$$\gamma(E_k) \leq e^{-Mk}$$

for k large. It follows that

$$\sum_{k=1}^{\infty} 2^{mk}\gamma(E_k) \tag{3.8}$$

converges for each fixed integer $m \geq 1$. According to the Melnikov-Hallstrom Theorem (cf. [4]), the convergence of the series (3.8) is necessary and sufficient for the differentiation functional $f \to f^{(m)}(p)$ to extend continuously to $R(K)$. This concludes the proof. □

It is easy to find an example of a set K such that $0 \in \partial K$ has a unique Jensen measure, while 0 does not have a unique representing measure for $R(K)$. Since such an example is required for the next chapter, we provide some details (cf. [3]).

Let K be the roadrunner set obtained from the closed unit disc by excising the open discs $\{|z-3^{-j}| < 8^{-j}\}$, $1 \leq j \leq \infty$. One checks that the measure $\frac{dz}{2\pi iz}$ on ∂K has finite variation, and it is a complex representing measure for 0. It follows [2, p.33] that 0 has a representing measure distinct from the point mass. On the other hand, the functions

$$f_j(z) = 8^{-j}/(z-3^{-j})$$

are bounded in modulus by one on K, while $|f_j'(0)| = (9/8)^j$ tends to $+\infty$. Since the differentiation functional is not continuous, 0 belongs to the Jensen boundary for $R(K)$, by Theorem 3.8. Of course, one can check directly that the Wiener series diverges, so that 0 is a regular boundary point for the Dirichlet problem.

References

1. Debiard, A. and Gaveau, B. Potential fin et algèbras de fonctions analytiques I, *J. Functional Analysis* 16 (1974), 289-304.

2. Gamelin, T.W. *Uniform Algebras*, Prentice-Hall, 1969.
3. Gamelin, T.W. and Rossi, H. Jensen measures and algebras of analytic functions, in *Function Algebras*, F. Birtel (ed.), Scott, Foresman and Co., 1966, pp.15-35.
4. Hallstrom, A. On bounded point derivations and analytic capacity, *J. Functional Analysis* 4 (1969), 153-165.
5. Tsuji, M. *Potential Theory in Modern Function Theory*, Maruzen, Tokyo, 1959.

4 · The corona problem for Riemann surfaces

This chapter is based on some unpublished work of B.Cole, that has been circulating since about 1970. Cole constructed an open Riemann surface for which the corona problem has a negative answer. The key idea of the construction is related to Jensen measures.

Let D be a set with an analytic structure, and suppose that the algebra $H^\infty(D)$ of bounded analytic functions on D separates the points of D. We can then regard D as a subset of the maximal ideal space $M(D)$ of $H^\infty(D)$, by identifying $z \in D$ with the ideal of functions that vanish at z. The corona problem asks whether D is dense in $M(D)$.

The corona problem can be rephrased in terms of the functions in $H^\infty(D)$. The density of D in $M(D)$ is equivalent to the validity of the following condition:

Whenever $\delta > 0$ and $f_1,\ldots,f_n \in H^\infty(D)$ satisfy $|f_1| + \ldots + |f_n| \geq \delta > 0$ on D, there exist $g_1,\ldots,g_n \in H^\infty(D)$ such that $f_1 g_1 + \ldots + f_n g_n = 1$. (4.1)

We will also consider the following stronger property that D might or might not enjoy:

For all $\delta > 0$ and $n \in Z_+$, there exist constants $C(n,\delta)$ such that whenever $f_1,\ldots,f_n \in H^\infty(D)$ satisfy $|f_j| \leq 1$, $1 \leq j \leq n$, and $|f_1| + \ldots + |f_n| \geq \delta$, then there are $g_1,\ldots,g_n \in H^\infty(D)$ satisfying $f_1 g_1 + \ldots + f_n g_n = 1$, and (4.2)

$$|g_j| \le C(n,\delta) , \quad 1 \le j \le n .$$

We will focus our attention on the property (4.2). For more information and references, see [1].

According to the corona theorem of L.Carleson, the open unit disc Δ is dense in $M(\Delta)$. Carleson actually proves that Δ has the property (4.2). In [1] it is shown that any finitely connected planar domain D has property (4.2), with constants $C_m(n,\delta)$ depending only on the number m of boundary components of D . The corona problem for planar domains can be reinterpreted as asking whether the $C_m(n,\delta)$ are bounded as $m \to \infty$. If so, then all planar domains have property (4.2), and the same constants will serve for all such domains. If not, then it is possible to construct a planar domain D that fails to be dense in $M(D)$.

Our purpose here is to prove Cole's Theorem to the effect that the analogous constants for finite bordered Riemann surfaces fail to be bounded.

4.1 Theorem. Let $0 < \delta < 1$, and let $M > 0$. Then there exists a finite bordered Riemann surface R , together with $f_1, f_2 \in H^\infty(R)$, such that

$$|f_1| \le 1 , \quad |f_2| \le 1 ; \tag{4.3}$$

$$|f_1| + |f_2| \ge \delta ; \tag{4.4}$$

if $g_1, g_2 \in H^\infty(R)$ satisfy $f_1 g_1 + f_2 g_2 = 1$, (4.5)
then
$\max(\|g_1\|_R, \|g_2\|_R) \ge M$.

Proof. We begin with the roadrunner set K constructed at the close of the preceding chapter. For this example, the Choquet boundary of $H(K)$ coincides with ∂K . It is an elementary fact (cf. reference [3] of Chapter 3) that every function in $C_R(\partial K)$ is then the restriction to ∂K of a function in $\mathcal{H}(K)$. On the other hand, K is constructed so that there is a representing measure ν for $0 \in \partial K$ with respect to the algebra $R(K)$, such that ν has no mass at 0 .

Denote the open disc $\{|z| < \rho\}$ by Δ_ρ . Since $\nu(\{0\}) = 0$, we can choose $\varepsilon > 0$ and $\rho > 0$ so small that

$$2M\varepsilon + M\nu(\Delta_\rho) < 1 . \tag{4.6}$$

Choose functions u_1 and u_2 in $\mathcal{H}(K)$ such that $u_1 < 0$ on K , $u_1 < \log \varepsilon$ on $K \backslash \Delta_\rho$, $u_2 < 0$ on K , $u_2(0) < \log \varepsilon$, and $\max(u_1, u_2) > \log \delta$. By Lemma 3.2, we can assume that $u_1 = \frac{1}{N} \log |f_1|$ and $u_2 = \frac{1}{N} \log |f_2|$, where f_1 and f_2 are analytic and nonzero in a neighbourhood of K , and N is an integer. This yields a neighbourhood U of K which is bounded by a finite number of closed analytic curves, and functions f_1, f_2 analytic on U that satisfy

$$|f_1|, |f_2| < 1 \quad \text{on} \quad U , \tag{4.7}$$

$$f_1, f_2 \quad \text{do not vanish on} \quad U , \tag{4.8}$$

$$|f_1|^{1/N} + |f_2|^{1/N} > \delta \quad \text{on} \quad U , \tag{4.9}$$

$$|f_1|^{1/N} < \varepsilon \quad \text{on} \quad U \backslash \Delta_\rho , \quad \text{and} \tag{4.10}$$

$$|f_2(0)|^{1/N} < \varepsilon . \tag{4.11}$$

Consider the set of points $(z_1, z_2, z_3) \in C^3$ that satisfy $z_3 \in U$, $z_1^N = f_1(z_3)$, and $z_2^N = f_2(z_3)$. On account of (4.8), the coordinate projection π_3 onto the third coordinate plane is an unramified N^2-sheeted covering of this surface over U . Let R be a component of this surface, so that π_3 is an unramified covering map of R onto U. Define analytic functions F_1 and F_2 on R by setting $F_j(z) = z_j$, $j = 1,2$. Then $F_j^N = f_j \circ \pi_3$, and from (4.7) and (4.9) we obtain $|F_1| < 1$, $|F_2| < 1$ and $|F_1| + |F_2| > \delta$ on R.

Suppose there exist $G_1, G_2 \in H^\infty(R)$ such that $|G_1| < M$, $|G_2| < M$ and $F_1G_1 + F_2G_2 = 1$ on R. We will complete the proof by showing that this leads to a contradiction.

Set $H = F_1G_1$. Then H is analytic on R, $|H| < M$, $|H| < M|F_1|$, and $|1-H| < M|F_2|$. Let h be the trace of H on U, that is, define $h(z_3)$ to be the average of the values of H on the fiber $\pi_3^{-1}(z_3) \cap R$. Then h is analytic on U, and

$$|h| \le M|f_1|^{1/N}, \tag{4.12}$$

$$|h-1| \le M|f_2|^{1/N}. \tag{4.13}$$

From (4.7), (4.10) and (4.12) we see that $|h| \le M$, while $|h| \le M\varepsilon$ off the ρ-neighbourhood of 0. Hence

$$|h(0)| = |\int h d\nu| \le |\int_{\Delta_\rho} h d\nu| + |\int_{K\setminus\Delta_\rho} h d\nu| \le M\nu(\Delta_\rho) + M\varepsilon .$$

On the other hand, from (4.11) and (4.13) we have

$$|h(0) - 1| \le M\varepsilon .$$

These two estimates for $h(0)$ contradict (4.6). □

4.2 Theorem (Cole). There exists an open Riemann surface R that is not dense in the maximal ideal space $M(R)$ of $H^\infty(R)$.

Proof. Fix $0 < \delta < 1$. According to Theorem 4.1, there is a sequence $\{R_m\}_{m=1}^{\infty}$ of finite bordered Riemann surfaces, together with analytic functions F_m and G_m on R_m that extend analytically across the border of R_m, such that

$$|F_m| \le 1 , \ |G_m| \le 1 \quad \text{on} \quad R_m ; \tag{4.14}$$

$$|F_m| + |G_m| \ge \delta \quad \text{on} \quad R_m ; \tag{4.15}$$

$$\text{if} \quad \phi_m, \psi_m \in H^{\infty}(R_m) \quad \text{satisfy} \quad \phi_m F_m + \psi_m G_m = 1 , \tag{4.16}$$

$$\text{then} \quad ||\phi_m||_{\infty} + ||\psi_m||_{\infty} \ge m .$$

We claim that there is an ambient (connected) Riemann surface R containing the R_m's and their borders as disjoint subsets, and functions $F,G \in H^{\infty}(R)$, such that

$$|F| \le 2 , \ |G| \le 2 \quad \text{on} \quad R ; \tag{4.17}$$

$$|F| + |G| \ge \delta/2 > 0 \quad \text{on} \quad R , \tag{4.18}$$

$$|F-F_m| < 1/m \quad \text{and} \quad |G-G_m| < 1/m \quad \text{on} \quad R_m . \tag{4.19}$$

Suppose for the moment that R, F and G have been constructed, and that $\phi,\psi \in H^{\infty}(R)$ satisfy

$$\phi F + \psi G = 1 . \tag{4.20}$$

On R_m we then obtain

$$\phi_m F_m + \psi_m G_m = 1 ,$$

where

$$\phi_m = \phi/[1 + \phi(F_m - F) + \psi(G_m - G)] ,$$

$$\psi_m = \psi/[1 + \phi(F_m - F) + \psi(G_m - G)] .$$

Since the norms of ϕ_m and ψ_m on R_m are bounded as m tends to ∞, we obtain a contradiction to (4.16) for large m. We conclude that the equation (4.20) is not solvable in $H^\infty(R)$, consequently R is not dense in $M(R)$. It remains to construct R, F and G.

First we construct an ambient surface S containing the R_m's as follows. Let S_m be a finite bordered surface obtained from R_m by attaching an annulus to each boundary component of $\bar{R}_m$. Then S is obtained by attaching, for each $m \geq 1$, one of the boundary components of S_m to one of those of S_{m+1} by means of a rectangular strip.

Let T_m be a simple analytic arc that starts at ∂R_m, passes directly through the rectangle joining S_m to S_{m+1}, and terminates at ∂R_{m+1}. We assume that the T_m's are disjoint. Let E be the union of the R_m's, their boundaries, and the T_m's. Then E is a closed subset of S, and E is connected.

Define continuous functions f and g on E such that for each $m \geq 1$, f coincides with F_m and g with G_m on R_m, and such that

$$|f| \leq 1 \quad \text{and} \quad |g| \leq 1 \quad \text{on} \quad E ,$$

$$|f| + |g| \geq \delta \quad \text{on} \quad E .$$

Let F and G be functions analytic on S such that

$$|F-f| < \min(\delta/4, 1/(2m)) \quad \text{on} \quad \bar{R}_m \cup T_m ,$$

$$|G-g| < \min(\delta/4, 1/(2m)) \quad \text{on} \quad \bar{R}_m \cup T_m .$$

Such F and G can be constructed by means of a simple

iteration argument, together with Bishop's Theorem asserting that if K is a compact subset of an open Riemann surface S such that $S\backslash K$ is connected, then every continuous function on K holomorphic on the interior of K can be approximated uniformly on K by functions holomorphic on S. The existence of F and G also follows from a version of Arakelyan's Theorem for Riemann surfaces due to S.Scheinberg [2]. In any event, with F and G in hand, we take for the open surface R a connected open neighbourhood of E, small enough so that (4.17), (4.18) and (4.19) are valid.

□

Cole's example can be modified to provide an example of a bounded domain of holomorphy in $\mathbb{C}^3$ for which the answer to the corona problem is negative. Indeed, by embedding the surface R_m in a polydisc in $\mathbb{C}^3$, and fattening it, one obtains a domain D_m in $\mathbb{C}^3$ with smooth strictly pseudoconvex boundary, and functions F_m and G_m in $H^\infty(D_m)$ that satisfy (4.14), (4.15) and (4.16) relative to D_m. By translating and dilating the D_m's, we can arrange that the D_m's have pairwise disjoint closures, while the D_m's converge to 0. It can be shown that the D_m's can be joined by smooth narrow tubes so as to obtain a domain D in $\mathbb{C}^3$ with the following properties:

(i) Every point of $(\partial D)\backslash\{0\}$ is a smooth strictly pseudoconvex boundary point for D.

(ii) D is holomorphically convex, and in fact D is a Runge domain.

(iii) D is not dense in $\mathcal{M}(D)$.

An example of a holomorphically convex domain D in $\mathbb{C}^2$ that is not dense in $\mathcal{M}(D)$ has been constructed more recently by N.Sibony[3]. Meanwhile, it is unknown whether the corona problem has an affirmative solution for the unit ball or the unit polydisc in $\mathbb{C}^2$.

References

1. Gamelin, T. Localization of the corona problem, *Pac. J. Math.* 34 (1970), 73-81.
2. Scheinberg, S. Uniform approximation by functions analytic on a Riemann surface, *Annals of Math.*, to appear.
3. Sibony, N. Prolongement analytique des fonctions holomorphes bornées, *C. R. Acad. Sci. Paris*, t. 275 (1972), 973-976.

5 · Subharmonicity with respect to a uniform algebra

Let A be a uniform algebra on a compact space X, and let M_A denote the maximal ideal space of A. In this chapter, we continue the line of investigation begun in Chapters 1 and 2. We will introduce and treat various classes of "quasi-subharmonic" functions. The lower semi-continuous, quasi-subharmonic functions will be the log-envelope functions introduced in Chapter 2. The upper semi-continuous, quasi-subharmonic functions will be called simply "subharmonic". The subharmonic functions in this context correspond to the subharmonic functions on an open subset of $\mathbb{C}$, or to the plurisubharmonic functions on an open subset of $\mathbb{C}^n$.

The main theorems of this chapter are Theorems 5.9 and 5.10, asserting that a locally subharmonic function is subharmonic, while a bounded, locally log-envelope function is a log-envelope function. Our exposition will be based on work of the author and N.Sibony[3,4].

Quasi-subharmonic Functions

Let u be a Borel function from a subset S of M_A to $[-\infty,+\infty]$. We say that u is *quasi-subharmonic on* S if $u(\phi) \le \int u d\sigma$ for all $\phi \in S$ and all Jensen measures σ for ϕ supported on a compact subset of S. It is understood implicitly that the negative part of u, $\min(u,0)$, is integrable with respect to the Jensen measures σ for those $\phi \in S$ satisfying $u(\phi) > -\infty$.

Evidently u is quasi-subharmonic on S if and only if u is quasi-subharmonic on each compact subset of S.

A function u from S to $[-\infty,+\infty)$ is *subharmonic* if u is upper semi-continuous and quasi-subharmonic. A subharmonic function on S is bounded above on each compact subset of S .

Recall from Chapter 2 that a lower semi-continuous function from a closed subset E of M_A to $(-\infty,+\infty]$ is a *log-envelope function* if it is the upper envelope of functions of the form $c \log|f|$, where $c > 0$ and $f \in A$. Some elementary properties of log-envelope functions are collected in (1.10) through (1.13).

5.1 Lemma. Let u be a lower semi-continuous function from a compact subset E of M_A to $(-\infty,+\infty]$. Then u is a log-envelope function if and only if u is quasi-subharmonic.

Proof. This is a special case of Theorem 1.6, in which $\mathcal{R}$ is the family of functions $c \log|f|$, where $c > 0$ and $f \in A$. □

In particular, a continuous function on E is subharmonic if and only if it is a log-envelope function. A compactness argument yields easily the following further characterization, which is a special case of Lemma 1.4.

5.2 Lemma. The following are equivalent, for a continuous, real-valued function u on a compact subset E of M_A .

(i) u is subharmonic on E .

(ii) u is a log-envelope function on E .

(iii) u is a uniform limit on E of functions of the form $\max(c_1\log|f_1|,\ldots,c_m\log|f_m|)$, where $c_1,\ldots,c_m$ are positive real numbers, and $f_1,\ldots,f_m \in A$.

In order to characterize the subharmonic functions, we recall some facts from Chapters 1 and 2 concerning the A-Dirichlet problem.

Suppose that u is a lower semi-continuous function, defined on a compact subset E of M_A , with values in

$(-\infty,+\infty]$. The *solution to the A-Dirichlet problem* with boundary data u on E is defined to be the upper envelope $\tilde{u}$ of all log-envelope functions on M_A that are dominated by u on E. The solution $\tilde{u}$ can be expressed in the form

$$\tilde{u}(\phi) = \sup\{c \log|f(\phi)| : c > 0, f \in A, c \log|f| < u \text{ on } E\}. \tag{5.1}$$

According to Edwards' Theorem, $\tilde{u}$ can be expressed in terms of Jensen measures,

$$\tilde{u}(\phi) = \inf\{\int u d\sigma : \sigma \text{ a Jensen measure on } E \text{ for } \phi\}. \tag{5.2}$$

Some elementary properties of the correspondence $u \to \tilde{u}$ are listed in (1.17) through (1.22).

5.3 Lemma. Let u be an upper semi-continuous function from a compact subset E of M_A to $[-\infty,+\infty)$. Then u is subharmonic on E if and only if u is the pointwise limit of a decreasing net of continuous subharmonic functions on E.

Proof. Suppose that u is the pointwise limit of a decreasing net $\{u_\alpha\}$ of subharmonic functions on E. Let $\phi \in E$, and let σ be a Jensen measure on E for ϕ. Suppose $v \in C_R(E)$ satisfies $u < v$. Then $u_\alpha < v$ for large α, so that $u(\phi) \le u_\alpha(\phi) \le \int u_\alpha d\sigma \le \int v d\sigma$. Taking the infimum over such v, we obtain $u(\phi) \le \int u d\sigma$. Hence u is subharmonic.

Conversely, suppose that u is subharmonic on E. It suffices to find, for each $w \in C_R(E)$ satisfying $u < w$, a continuous subharmonic function v on E such that $u < v < w$. For this, choose $\varepsilon > 0$ such that $u < w - \varepsilon$. Let $\phi \in E$. If σ is a Jensen measure on E for ϕ, then $u(\phi) \le \int u d\sigma \le \int w d\sigma - \varepsilon$. Taking the infimum over such σ, we obtain $u(\phi) \le \tilde{w}(\phi) - \varepsilon$. By Edwards' Theorem, there exist $c > 0$ and $f \in A$ such that $c \log|f| < w$ on E, while $u(\phi) < c \log|f(\phi)|$. The latter inequality persists in a neighbourhood of ϕ. Covering E by a finite number of such neighbourhoods, and letting $c_j \log|f_j|$ be the corresponding functions, we find that $v = \max c_j \log|f_j|$ is a continuous subharmonic function that satisfies $u < v < w$. □

Let U be an open subset of M_A. A function u from U to $[-\infty,+\infty)$ is *locally subharmonic* if each point of U has a compact neighbourhood N such that $u|_N$ is subharmonic on N. The *locally log-envelope functions* and the *locally quasi-subharmonic functions* on U are defined similarly. In order to study these classes of functions, we take a detour through algebras generated by Hartogs series. This will lead to a further characterization of log-envelope functions.

Hartogs Series

A Hartogs series is a series of the form $\Sigma f_j \zeta^j$, where ζ is a complex variable, and the f_j's depend on other parameters. We are interested in the case that the f_j's belong to the algebra A.

Let u be a lower semi-continuous function from X to $(-\infty,+\infty]$. Define a subset Y of $X \times \mathbb{C}$ by

$$Y = \{(x,\zeta) : x \in X, \ |\zeta| \le e^{-u(x)}\} . \tag{5.3}$$

The lower semi-continuity of u guarantees that Y is compact.

We regard A as a subalgebra of $C(Y)$ in the obvious way, and we define B to be the uniform algebra on Y generated by A and the coordinate function ζ. A dense subalgebra of B is formed by the Hartogs polynomials

$$F(x,\zeta) = \sum_{j=0}^{N} f_j(x)\zeta^j \ , \quad (x,\zeta) \in Y \ , \tag{5.4}$$

where $f_0,\ldots,f_N \in A$. Since each $\psi \in M_B$ is determined by its restriction to A and its value on the coordinate function ζ, any such ψ has the form

$$\psi(F) = \sum_{j=0}^{N} f_j(\phi)\zeta_0^j \ ,$$

where $\phi = \psi|_A$, $\zeta_0 = \psi(\zeta)$, and F is as in (5.4). Consequently M_B can be regarded as a compact subset of $M_A \times \mathbb{C}$.

There is another way to regard M_B. Let $A \otimes P$ denote the algebra of Hartogs polynomials on $M_A \times \mathbb{C}$, that is, the algebra of functions of the form

$$F = \sum_{j=0}^{N} f_j\zeta^j \ , \quad f_0,\ldots,f_N \in A \ . \tag{5.5}$$

The $A \otimes P$-convex hull $\hat{E}$ of a compact subset E of $M_A \times \mathbb{C}$ consists by definition of those $(\phi,\zeta_0) \in M_A \times \mathbb{C}$ such that the evaluation functional $F \to F(\phi,\zeta_0)$ is continuous with respect to the norm of uniform convergence on E. This occurs if and only if

$$|F(\phi,\zeta_0)| \le \|F\|_E \ , \quad F \in A \otimes P \ .$$

Then $\hat{E}$ is easily seen to be the maximal ideal space of the uniform closure in $C(E)$ of $A \otimes P$. In particular, M_B coincides with the $A \otimes P$-convex hull $\hat{Y}$ of Y.

Since the A-convex hull of X is M_A, the $A \otimes P$-convex hull of $X \times \{0\}$ coincides with $M_A \times \{0\}$, and $M_A \times \{0\} \subseteq M_B$.

Now the algebra of Hartogs polynomials is invariant under rotations in the ζ-coordinate, and so is Y, so that M_B is invariant under rotations in the ζ-coordinate. Furthermore, the Hartogs polynomials depend analytically on the ζ

variable, so that the maximum modulus principle shows that M_B includes the entire disc $\{(\phi,\zeta) : |\zeta| \le r\}$ just as soon as it contains the boundary circle $\{(\phi,\zeta) : |\zeta| = r\}$. It follows that M_B has the form

$$M_B = \{(\phi,\zeta) : |\zeta| \le R(\phi)\} ,$$

where R is some function from M_A to $[0,+\infty)$. Since M_B is compact, R is bounded and upper semi-continuous.

How can R be captured from the function u? The answer turns out to be reasonably simple: $R = \exp(-\tilde{u})$. We state the result in full.

5.4 Theorem. Let u be a lower semi-continuous function from X to $(-\infty,+\infty]$, let Y be as in (5.3), and let B be the uniform closure in $C(Y)$ of the algebra $A \otimes P$ of Hartogs polynomials. Then

$$M_B = \{(\phi,\zeta) : \phi \in M_A , |\zeta| \le e^{-\tilde{u}(\phi)}\} , \tag{5.6}$$

where $\tilde{u}$ is the solution to the A-Dirichlet problem with boundary function u.

Proof. Suppose that $|\zeta_0| > \exp(-\tilde{u}(\phi))$, that is, that $-\log|\zeta_0| < \tilde{u}(\phi)$. Then there exist $f \in A$ and a positive integer m such that $\frac{1}{m}\log|f| < u$ on X, while $-\log|\zeta_0| < \frac{1}{m}\log|f(\phi)|$. The first estimate shows that $|f|\exp(-mu) \le 1$ on X, so that $|\zeta^m f| \le 1$ on Y. The second estimate shows that $|\zeta_0^m f(\phi)| > 1$. Hence (ϕ,ζ_0) does not belong to $\hat{Y} = M_B$.

It suffices now to show that $(\phi,\zeta_0) \in M_B$ whenever $|\zeta_0| \le \exp(-\tilde{u}(\phi))$. Since M_B includes $M_A \times \{0\}$, we may assume that $\zeta_0 \ne 0$, so that in particular $\tilde{u}(\phi) < +\infty$.

Suppose first that $|\zeta_0| < \exp(-\tilde{u}(\phi))$. Let F be a

Hartogs polynomial of the form (5.5), and suppose that $|F| \leq 1$ on Y. For fixed $x \in X$, the polynomial $\Sigma f_j(x)\zeta^j$ is bounded in modulus by 1 on the disc $|\zeta| \leq \exp(-u(x))$. The Cauchy estimates for the coefficients $f_j(x)$ then take the form $|f_j(x)| \leq \exp(ju(x))$, so that $\log|f_j| \leq ju$ on X. It follows that $\log|f_j| \leq j\tilde{u}$ on M_A. Consequently $|f_j(\phi)| \leq \exp(j\tilde{u}(\phi))$, and

$$
\begin{aligned}
|F(\phi,\zeta_0)| &\leq \sum_{j=0}^{N} |f_j(\phi)||\zeta_0|^j \\
&\leq \sum_{j=0}^{N} |\zeta_0 \, e^{\tilde{u}(\phi)}|^j \\
&\leq 1/[1 - |\zeta_0| \, e^{\tilde{u}(\phi)}] .
\end{aligned}
$$

Since this estimate is independent of N, the evaluation functional $F \to F(\phi,\zeta_0)$ extends continuously to B. Hence (ϕ,ζ_0) belongs to M_B, this for all ζ_0 satisfying $|\zeta_0| < \exp(-\tilde{u}(\phi))$. Since M_B is compact, M_B also includes (ϕ,ζ_0) whenever $|\zeta_0| = \exp(-\tilde{u}(\phi))$. □

As a corollary, we obtain a characterization of log-envelope functions.

5.5 Corollary. Let w be a lower semi-continuous function from M_A to $(-\infty,+\infty]$. Then w is A-subharmonic if and only if the set

$$\{(\phi,\zeta) : \phi \in M_A , |\zeta| \leq e^{-w(\phi)}\} \tag{5.7}$$

is an $A \otimes P$-convex subset of $M_A \times \mathbb{C}$.

Proof. Applying Theorem 5.4 in the case $X = M_A$, we find that the $A \otimes P$-convex hull of the set defined by (5.7) is given by

$$\{(\phi,\zeta) : \phi \in M_A , |\zeta| \le e^{-\tilde{w}(\phi)}\} .$$

Consequently the set defined by (5.7) is $A \otimes P$-convex if and only if $w = \tilde{w}$. □

Algebras Generated by Hartogs-Laurent Series

There are analogous results for algebras generated by Hartogs-Laurent polynomials of the form

$$F = \sum_{j=-N}^{N} f_j \zeta^j , \tag{5.8}$$

where the f_j's belong to A . Let u and v be lower semi-continuous functions such that $v \le -u$, and define a compact subset Y of $X \times \mathbb{C}$ by

$$Y = \{(x,\zeta) : e^{v(x)} \le |\zeta| \le e^{-u(x)}\} .$$

Let B be the uniform closure in $C(Y)$ of functions of the form (5.8), so that B is generated by A , ζ and $1/\zeta$.

Since every complex-valued homomorphism of B is determined by its action on A and on ζ , the maximal ideal space M_B of B can again be identified with a compact subset of $M_A \times \mathbb{C}$. It consists of all pairs (ϕ,ζ) such that the evaluation functional $F \to F(\phi,\zeta)$, defined on Hartogs-Laurent polynomials F , extends continuously to B . The space M_B is described directly in terms of u and v as follows.

5.6 Theorem. Let u , v , Y and B be as above. Then

$$M_B = \{(\phi,\zeta) : \phi \in M_A , e^{\tilde{v}(\phi)} \le |\zeta| \le e^{-\tilde{u}(\phi)}\} , \tag{5.9}$$

where $\tilde{u}$ and $\tilde{v}$ are the solutions to the A-Dirichlet problem associated with u and v respectively.

Proof. As in the preceding proof, it is easy to check that M_B is included in the set described by (5.9).

Suppose that (ϕ,ζ_0) satisfies

$$e^{\tilde{v}(\phi)} < |\zeta_0| < e^{-\tilde{u}(\phi)} . \tag{5.10}$$

Let F be a Hartogs-Laurent polynomial of the form (5.8), and suppose that $|F| \le 1$ on Y. The coefficient functions f_j are given by

$$f_j(x) = \frac{1}{2\pi i}\int_{|\zeta|=c} F(x,\zeta)\zeta^{-j-1}d\zeta , \quad x \in X . \tag{5.11}$$

Suppose that $j \ge 0$. Taking $c = \exp(-u(x))$ in (5.11), we are led to the estimate $|f_j(x)| \le \exp(ju(x))$, or $\log|f_j| \le ju$ on X. Hence $\log|f_j| \le j\tilde{u}$ on M_A, and $|f_j(\phi)| \le \exp(j\tilde{u}(\phi))$. This yields the estimate

$$\left|\sum_{j=0}^{N} f_j(\phi)\zeta_0^j\right| \le 1/[1-|\zeta_0|e^{\tilde{u}(\phi)}] . \tag{5.12}$$

On the other hand, suppose that $j<0$. Setting $c=\exp(v(x))$ in (5.11), we obtain $|f_j(x)| \le \exp(-jv(x))$, or $\log|f_j| \le -jv$ on X. Hence $\log|f_j| \le -j\tilde{v}$ on M_A, and $|f_j(\phi)| \le \exp(-j\tilde{v}(\phi))$. This yields the estimate

$$\left|\sum_{j=-N}^{-1} f_j(\phi)\zeta_0^j\right| \le 1/[1-|\zeta_0|^{-1} e^{\tilde{v}(\phi)}] . \tag{5.13}$$

Combining (5.12) and (5.13), we obtain a bound for $|F(\phi,\zeta_0)|$ which is independent of N. It follows that the functional $F \to F(\phi,\zeta_0)$ extends continuously to B. Hence M_B includes all (ϕ,ζ_0) that satisfy (5.10). Since M_B is closed, M_B includes all (ϕ,ζ) that satisfy $\exp(\tilde{v}(\phi)) \le |\zeta| \le \exp(-\tilde{u}(\phi))$, providing that $\tilde{v}(\phi) < -\tilde{u}(\phi)$. This proves the theorem whenever $\tilde{v} < -\tilde{u}$ on M_A.

For the general case, let $\varepsilon > 0$, define $v_\varepsilon = v-\varepsilon$, set

$$Y_\varepsilon = \{(x,\zeta) : x \in X,\ e^{v(x)-\varepsilon} \le |\zeta| \le e^{-u(x)}\},$$

and let B_ε be the uniform closure on Y_ε of the Hartogs-Laurent polynomials. Since the Y_ε decrease to Y as ε decreases to 0, also M_{B_ε} decreases to M_B as ε decreases to 0. Now observe that since $u+v_\varepsilon \le -\varepsilon$, also $\tilde{u}+\tilde{v}_\varepsilon \le \widetilde{u+v}_\varepsilon \le -\varepsilon$. Hence $\tilde{v}_\varepsilon < -\tilde{u}$ on M_A, and the special case of the theorem already established applies to B_ε. Since $\tilde{v}_\varepsilon = \tilde{v}-\varepsilon$, we obtain

$$M_{B_\varepsilon} = \{(\phi,\zeta) : \phi \in M_A,\ e^{\tilde{v}(\phi)-\varepsilon} \le |\zeta| \le e^{-\tilde{u}(\phi)}\}.$$

Letting ε decrease to zero, we obtain (5.9). □

Since $u+v \le 0$, also $\tilde{u}+\tilde{v} \le 0$, and $\tilde{v} \le -\tilde{u}$ on M_A. The slices of the set (5.9) corresponding to fixed values of ϕ are therefore either circles or annuli, never empty. In case the slice is an annulus, the functions in B depend analytically on ζ inside the annulus.

Consider now the Laurent series of a function F in B, for a fixed $\phi \in M_A$:

$$F(\phi,\zeta) \sim \sum_{j=-\infty}^{j=+\infty} f_j(\phi)\zeta^j .$$

The coefficient $f_j(\phi)$ is given by

$$f_j(\phi) = \frac{1}{2\pi i}\int_{|\zeta|=c} F(\phi,\zeta)\zeta^{-j-1}d\zeta ,$$

where c is any fixed number in the range $\exp(\tilde{v}(\phi)) \le c \le \exp(-\tilde{u}(\phi))$. If F is the Hartogs-Laurent polynomial given by (5.8), then the coefficients f_j of F occurring in (5.8) coincide with those defined above. The integral formula makes it clear that the coefficient functions f_j depend continuously on F, in the norm of B. It follows that the coefficient function f_j belongs to A for each $F \in B$. One consequence of this observation is that a Laurent polynomial $\Sigma_{j=-N}^{N} g_j(\phi)\zeta^j$ belongs to B only when each of the coefficient functions g_j belongs to A. In particular, B is a proper subalgebra of $C(Y)$ just as soon as A is a proper subalgebra of $C(X)$.

As an illustration of this class of algebras, we mention an example due to R.Basener, of a uniform algebra B such that $B \neq C(M_B)$, while every point of M_B is a peak point for B. The first such example had been discovered by B. Cole[1, p.255], and Cole's construction also depends crucially on the notion of Jensen measure.

For Basener's example, we start with a compact subset K of the open unit disc in $\mathbb{C}$ such that $R(K) \neq C(K)$, while every point of K has a unique Jensen measure (namely, the point mass) with respect to the algebra $R(K)$. For such a K, one can take J.Wermer's example [5] of a compact subset K of $\mathbb{C}$ such that $R(K)$ has no continuous point derivations. (Recall that by Theorem 3.8, point derivations of all orders exist at any point for which the point mass is not a unique Jensen measure.) One can also take K to be McKissick's celebrated Swiss cheese. In any event, we take $R(K)$ to be the algebra A in the preceding discussion, and we define

$$u(z) = -v(z) = -\log\sqrt{1 - |z|^2}, \quad z \in K.$$

The set Y of the preceding theorem becomes the compact subset

$$Y = \{(z, e^{i\theta}\sqrt{1 - |z|^2}) : z \in K, \ 0 \le \theta \le 2\pi\}$$

of $\mathbb{C}^2$. Since Jensen measures are unique, we have $\tilde{u} = -\tilde{v} = u = -v$, and the preceding theorem shows that M_B coincides with Y . According to our earlier remarks, $B \neq C(M_B)$. On the other hand, note that Y lies on the unit sphere of $\mathbb{C}^2$, so that every point of M_B is a peak point for B . In fact, the peaking functions can be taken to be a linear combination of z and ζ .

Localization Principle for Subharmonic Functions

Before beginning, we introduce two pieces of notation that will be useful.

If E is a closed subset of M_A , then A_E will denote the uniform closure in $C(E)$ of the restriction algebra $A|_E$. The maximal ideal space of A_E can be identified with the A-convex hull $\hat{E}$ of E . By Rossi's local maximum modulus principle, the Shilov boundary of A_E is included in $(\partial E) \cup (X \cap E)$.

For the second piece of notation, we will denote by u^* the upper semi-continuous regularization of a function u on a set S , defined by

$$u^*(y) = \limsup_{S \ni s \to y} u(s) ,$$

for y belonging to the closure of S .

The following lemma provides the key to the localization of subharmonicity.

5.7 Lemma. Let u be a lower semi-continuous function from X to $(-\infty,+\infty]$. Let E be a compact subset of M_A , and let v be a quasi-subharmonic function on E . If $v \leq u$ on $X \cap E^o$, while $v \leq \tilde{u}$ on ∂E , then $v \leq \tilde{u}$ on E.

Proof. Define a lower semi-continuous function w on $(X \cap E^o) \cup \partial E$ by

$$w(\phi) = \begin{cases} u(\phi) , & \phi \in X \cap E^o , \\ \tilde{u}(\phi) , & \phi \in \partial E . \end{cases}$$

Let $\tilde{w}$ be the solution to the A-Dirichlet problem with boundary data w on $(X \cap E^o) \cup \partial E$. First we show that $\tilde{w} \le \tilde{u}$ on E.

Let Y be defined as in (5.3), so that the maximal ideal space M_B of the uniform closure B of $A \otimes P$ in $C(Y)$ is given by (5.7). Let $\Pi : M_B \to M_A$ be the natural projection, and define

$$Q = [Y \cap \Pi^{-1}(E^o)] \cup \Pi^{-1}(\partial E) .$$

Note that Q includes the boundary of $\Pi^{-1}(E)$, and Q also includes the intersection of $\Pi^{-1}(E)$ with the Shilov boundary of B. By Rossi's local maximum modulus principle, Q includes the Shilov boundary of the restriction algebra $B_{\Pi^{-1}(E)}$. In other words, the $A \otimes P$-convex hull of Q includes $\Pi^{-1}(E)$. In particular, if $\phi \in E$, then $(\phi, \exp(-\tilde{u}(\phi)))$ belongs to the $A \otimes P$-convex hull of Q.

Now observe by inspection that

$$Q = \{(\phi,\zeta) : \phi \in (X \cap E^o) \cup \partial E , |\zeta| \le e^{-w(\phi)}\} .$$

By Theorem 5.4, applied with A_E in place of A, Q in place of Y and w in place of u, any (ϕ,ζ) lying in the $A_E \otimes P$-convex hull of Q satisfies $|\zeta| \le \exp(-\tilde{w}(\phi))$. Now the $A_E \otimes P$-convex hull of Q coincides with the $A \otimes P$-convex hull. We obtain then $\exp(-\tilde{u}(\phi)) \le \exp(-\tilde{w}(\phi))$ for $\phi \in E$, and $\tilde{w} \le \tilde{u}$ on E.

By hypothesis, $v \le w$ on $(X \cap E^o) \cup \partial E$. If $\phi \in E$, and σ is a Jensen measure on $(X \cap E^o) \cup \partial E$ for ϕ, then $v(\phi) \le \int v d\sigma \le \int w d\sigma$. By Edwards' Theorem, the infimum over all such σ of $\int w d\sigma$ coincides with $\tilde{w}(\phi)$. Hence we obtain $v(\phi) \le \tilde{w}(\phi)$, and $v \le \tilde{w} \le \tilde{u}$ on E. □

The next lemma is a version of the maximum principle.

5.8 Lemma. Let U be an open subset of M_A such that every point of ∂U is a Jensen boundary point for A. Let u be a lower semi-continuous function from X to $(-\infty,+\infty]$. Let v be a locally quasi-subharmonic function on U that is bounded above. If $v \le u$ on $X \cap U$, while $v^* \le u$ on ∂U, then $v \le \tilde{u}$ on U.

Proof. Note first that the Jensen boundary is included in X, so that u is defined on $\partial U \subseteq X$.

For the purposes of obtaining a contradiction, suppose that

$$\alpha = \sup\{v(\phi) - \tilde{u}(\phi) : \phi \in U\}$$

is strictly positive. Since v is bounded above, α is finite. Define

$$E = \{\phi \in U : \limsup_{y \to \phi} [v(\phi) - \tilde{u}(\phi)] = \alpha\} .$$

Since points of ∂U are Jensen boundary points, $\tilde{u} = u$ on ∂U. The hypothesis $v^* \le u$ on ∂U then implies that E does not reach out to ∂U, so that E is a compact subset of U.

Let ϕ_o belong to the Shilov boundary of A_E. By hypothesis, there is a compact neighbourhood N of ϕ_o on which v is quasi-subharmonic. We may assume that $N \subset U$.

Choose $g \in A$ such that

$$\sup\{|g(\phi)| : \phi \in E\} = 2 ,$$

while

$$|g| < 1/2 \quad \text{on} \quad E\backslash N^o .$$

This choice is possible, since the Shilov boundary of A_E meets N^o. The choice of g and the definition of α and E show that for $c > 0$ sufficiently large,

$$c(v-\tilde{u}-\alpha) + \log|g| < 0 \quad \text{on} \quad \partial N .$$

We choose c so large that also

$$-c\alpha + \log\|g\| < 0 ,$$

the norm being that of A. The latter estimate, together with the hypothesis $v \le u$ on $X \cap U$, shows that

$$c(v-\alpha) + \log|g| < cu \quad \text{on} \quad X \cap N ,$$

whilst the former estimate shows that

$$c(v-\alpha) + \log|g| < c\tilde{u} \quad \text{on} \quad \partial N .$$

Now $c(v-\alpha) + \log|g|$ is quasi-subharmonic on N. Applying Lemma 5.7, we find that $c(v-\alpha) + \log|g| \le c\tilde{u}$ on N, or

$$c(v-\tilde{u}-\alpha) + \log|g| \le 0 \quad \text{on} \quad N . \tag{5.14}$$

Now choose $\phi_1 \in E \cap N^o$ such that $|g(\phi_1)| = 2$. Then

$$\limsup_{y \to \phi_1} [c(v(y) - \tilde{u}(y) - \alpha) + \log|g(y)|] = \log 2 ,$$

and this contradicts (5.14). □

This concludes the uphill stretch. Now we can coast through the localization theorems and their corollaries.

5.9 Theorem. Let v be an upper semi-continuous function from M_A to $[-\infty,+\infty)$. Let U be an open subset of M_A such that v is locally subharmonic on U, while every point of $M_A \backslash U$ is a Jensen boundary point. Then v is subharmonic on M_A.

Proof. Let σ be a Jensen measure for $\phi \in M_A$. Let $u \in C_R(M_A)$ satisfy $u \geq v$. Then $v \leq \tilde{u}$ on $M_A \backslash U$, since u coincides with $\tilde{u}$ on the Jensen boundary. On U we obtain $v \leq \tilde{u}$ from Lemma 5.8. Hence $v(\phi) \leq \tilde{u}(\phi) \leq \int \tilde{u} d\sigma \leq \int u d\sigma$. Taking the infimum over such u, we obtain $v(\phi) \leq \int v d\sigma$, so that v is subharmonic. □

The exact analogue of Theorem 5.9 for log-envelope functions is false. Suppose for instance that each point of X is a Jensen boundary point for A, while $U = M_A \backslash X$ is not empty. Then the characteristic function of U is a lower semi-continuous function on M_A that is a locally log-envelope function on U, but is not a log-envelope function on M_A. However, there is a slightly more complicated version of the localization theorem that is valid for log-envelope functions.

5.10 Theorem. Let U be an open subset of M_A such that each point of $M_A \backslash U$ is a Jensen boundary point. Let u be a bounded, lower semi-continuous function on M_A such that u is a locally log-envelope function on U, while u is continuous at each point of $M_A \backslash U$. Then u is a log-envelope function on M_A.

Proof. Since each point of $M_A \backslash U$ is a Jensen boundary point,

$u = \tilde{u}$ on $M_A \backslash U$. Now apply Lemma 5.8 to $v = u$, to conclude that $u \leq \tilde{u}$ on U . Hence $u = \tilde{u}$, and u is a log-envelope function. □

5.11 Corollary. A locally subharmonic function on M_A is subharmonic. A bounded, locally log-envelope function on M_A is a log-envelope function.

Applying the preceding corollary to the algebra A_E , where E is an A-convex subset of M_A , we obtain the following.

5.12 Corollary. Let U be an open subset of M_A , and let E be a compact, A-convex subset of U . Then any locally subharmonic function on U is subharmonic on E . Any locally bounded, locally log-envelope function on U is a log-envelope function on E .

Try to prove the following theorem from the definitions!

5.13 Theorem. Let U be an open subset of M_A . The pointwise limit of a decreasing net of locally subharmonic functions on U is locally subharmonic. The pointwise limit of an increasing net of bounded, locally log-envelope functions on U is a locally log-envelope function.

Proof. Let u be the pointwise limit of the decreasing net $\{u_\alpha\}$ of locally subharmonic functions on U . Let $p \in U$. Choose a compact, A-convex neighbourhood N of p contained in U . By Corollary 5.12, each u_α is subharmonic on N . The proof of Lemma 5.3 shows that a pointwise limit of a decreasing net of subharmonic functions is again subharmonic. Hence u is subharmonic on N , and u is locally subharmonic on U .

The proof of the second statement is similar, once one observes that the pointwise limit of an increasing net of log-envelope functions is a log-envelope function. □

5.14 Theorem. Suppose that every point of X is a Jensen boundary point. Then any locally subharmonic function on $M_A \backslash X$ is subharmonic. Furthermore, the restrictions to $M_A \backslash X$ of the subharmonic functions on M_A are precisely the subharmonic functions on $M_A \backslash X$ that are bounded above.

Proof. Let E be a compact subset of $M_A \backslash X$. Then any $p \in \hat{E}$ has a Jensen measure supported on E. Consequently $\hat{E}$ does not meet X, and $\hat{E}$ is a compact subset of $M_A \backslash X$. Hence $M_A \backslash X$ is A-convex, and the first statement of the theorem follows easily from Corollary 5.12.

The restriction to $M_A \backslash X$ of a subharmonic function on M_A is subharmonic and bounded above. Conversely, if u is a subharmonic function on $M_A \backslash X$ that is bounded above, then u^* is an upper semi-continuous extension of u to M_A. Furthermore, u^* is subharmonic, by Theorem 5.9. □

Localization of the Jensen Boundary

Let $x_o \in X$, and let N be a compact neighbourhood of x_o in M_A. Rossi's theorem on local peak points asserts that if x_o is a peak point for the restriction algebra $A|_N$, then x_o is a peak point for A. However, it remains an open problem to determine whether x_o is a peak point for A under the weaker hypothesis that x_o be a peak point for the *closed* restriction algebra A_N.

The problem can be reformulated in terms of representing measures. Suppose that x_o has a compact neighbourhood N such that the point mass at x_o is the only representing measure for x_o on N. Is the point mass a unique representing measure for x_o ?

The analogous problem for Jensen boundary points can be solved on the basis of the localization theorem.

5.15 Theorem. Let A be a uniform algebra on X, and let $x_0 \in X$. Suppose there is a compact neighbourhood N of x_0 in M_A such that x_0 belongs to the Jensen boundary of A_N. Then x_0 belongs to the Jensen boundary of A.

Proof. Let $u \in C_R(M_A)$ satisfy $0 \le u \le 1$, while $u(x_0) = 1$, and $u = 0$ on $M_A \backslash N$. Let $\varepsilon > 0$. Since the point mass at x_0 is the unique Jensen measure for x_0 with respect to A_N, there exist $c > 0$ and $f \in A_N$ such that $c \log|f| < u$ on N, while $c \log|f(x_0)| > 1-\varepsilon$. We may assume that f belongs to A. Define a function v on M_A by setting $v = c \log^+|f|$ on N, and $v = 0$ off N. Then v is continuous, $v \le u$, and v vanishes off a compact subset of the interior of N. Clearly v is locally subharmonic. By Theorem 5.9, v is subharmonic.

If σ is any Jensen measure on M_A for x_0, then $\int u d\sigma \ge \int v d\sigma \ge v(x_0) \ge 1-\varepsilon$. Since $\varepsilon > 0$ is arbitrary, we obtain $\int u d\sigma = 1$, and σ is supported on the set $\{u=1\}$. On account of the freedom of choice of u, σ is the point mass at x_0. □

One corollary to the preceding theorem is that if ϕ is not a Jensen boundary point, then any subharmonic function u on a neighbourhood of ϕ satisfies

$$u(\phi) = \limsup_{y \ne \phi, y \to \phi} u(y) .$$

Indeed, on the basis of Theorem 5.15, there is a net $\{\sigma_\alpha\}$ of Jensen measures for ϕ such that the supports of the σ_α's tend to $\{\phi\}$, while each σ_α has no mass at ϕ. The estimate $u(\phi) \le \int u d\sigma_\alpha$ then implies that $u(\phi) \le \lim \sup u(y)$. The reverse inequality is the upper semicontinuity of u at ϕ.

Supports of Jensen Measures

As another immediate application of the localization theorem, we mention the following result.

5.16 Theorem. Let U be an open subset of M_A such that every point of ∂U is a Jensen boundary point. Then all Jensen measures for points of U are supported on $\bar{U}$.

Proof. Define $u = 1$ on $\bar{U}$ and $u = 0$ on $M_A \backslash U$. Then u is upper semi-continuous, and u is locally subharmonic on $M_A \backslash \partial U$. By Theorem 5.9, u is subharmonic on M_A. If σ is a Jensen measure for a point $\phi \in U$, then the estimate $1 = u(\phi) \leq \int u d\sigma$ shows that σ assigns full mass to $\bar{U}$.

□

It is proved in [4] that the Jensen measures for $\phi \in U$ supported on a compact subset of U are weak-star dense in the set of all Jensen measures for ϕ.

Theorem 5.16 is interesting, in part, because the corresponding assertion for representing measures is false. To see this, consider the "string of beads" example of A.M.Davie and J.Garnett (cf.[2,p.122]), with a disconnected Gleason part. In this example, K is a compact subset of the complex plane, every point of ∂K is a peak point for $R(K)$, and K^o consists of two connected components which together form a single Gleason part for $R(K)$. Then points of each component of K^o have representing measures with positive mass on the other component of K^o.

Bremermann Functions

We wish now to introduce a class of functions, the "Bremermann functions", which appear as solutions to a generalized Dirichlet problem. The class of functions will be defined locally. In one complex variable, these functions will correspond to the harmonic functions. In several complex variables, they will correspond to the plurisubharmonic functions whose complex Hessian matrices have determinant zero.

Let U be an open subset of $M_A\backslash X$. A lower semi-continuous function u on U is a *Bremermann function* if each $\phi \in U$ has a compact neighbourhood $N \subset U$ such that $u = \tilde{v}$ on N, where $v = u|_{\partial N}$. This latter condition is equivalent to the identity

$$u(\phi) = \inf_{\sigma} \int_{\partial N} u d\sigma, \quad \phi \in N, \tag{5.15}$$

where the infimum is taken over all Jensen measures σ for ϕ supported on ∂N. Note that the hypothesis $U \subset M_A\backslash X$ guarantees, by Rossi's local maximum modulus principle, that every $f \in A$ assumes its maximum modulus over N on ∂N. Consequently each $\phi \in N$ has a Jensen measure on ∂N, so that the infimum in (5.15) is well-defined.

Note that every Bremermann function is a locally log-envelope function.

The next two results serve to generate a number of Bremermann functions.

5.17 Lemma. Let E be a compact subset of M_A, and let u be a lower semi-continuous function from E to $(-\infty, +\infty]$. Then $\tilde{u}$ is a Bremermann function on $M_A\backslash(X \cup E)$.

Proof. Let $\phi \in M_A\backslash(X \cup E)$, and let N be any compact neighbourhood of ϕ disjoint from $X \cup E$. Then $\tilde{u}$ is a log-envelope function on N. Suppose that $c > 0$ and $f \in A$ satisfy $c \log|f| < \tilde{u}$ on ∂N. Let v be a continuous subharmonic function on M_A such that $v < \tilde{u}$, while $c \log|f| < v$ on ∂N. Define w on M_A so that $w = v$ on $M_A\backslash N$, while $w = \max(v, c \log|f|)$ on N. Then

w is continuous, and w is locally a log-envelope function. By Theorem 5.10, w is a log-envelope function. Since $w \le \tilde{u}$ on E, also $w \le \tilde{u}$ on N. Hence $c \log|f| \le \tilde{u}$ on N. It follows that $\tilde{u}$ is a Bremermann function. □

5.18 Lemma. If $c > 0$, $d \in \mathbb{R}$, $g \in A$ is invertible, and u is a Bremermann function on U, then $cu + d \log|g|$ is a Bremermann function on U. In particular, $\log|g|$ and $\mathrm{Re}(f)$ are Bremermann functions on $M_A \backslash X$ for all $f \in A$ and all invertible $g \in A$.

Proof. If σ is any Jensen measure for ϕ, then by applying the Jensen-Hartogs inequality to both g and $1/g$, we obtain $\log|g(\phi)| = \int \log|g| d\sigma$. It follows easily that the validity of (5.14) for u implies its validity for $cu + d \log|g|$, so that $cu + d \log|g|$ is a Bremermann function whenever u is. The final statement of the theorem follows from the first statement, upon setting $g = \exp(f)$. □

The maximum of two Bremermann functions need not be a Bremermann function, not even in the complex plane. Worse yet, the sum of two Bremermann functions need not be a Bremermann function. To see this, we anticipate some results of Chapter 6. It turns out that $|z_1|^2$ and $|z_2|^2$ are Bremermann functions on $\mathbb{C}^2$, since they are plurisubharmonic and their complex Hessian matrices are degenerate. However, $|z_1|^2 + |z_2|^2$ is not a Bremermann function, since its complex Hessian matrix is strictly positive. One can also see that $|z_1|^2 + |z_2|^2$ is not a Bremermann function by observing, from the expression (5.14), that a Bremermann function cannot attain a strict local minimum.

The main result on Bremermann functions is the following version of the maximum principle. The proof proceeds along the same lines as that of the maximum principle in Lemma 5.8.

5.19 Theorem. Let U be an open subset of $M_A \backslash X$. Let u be a Bremermann function on U that is bounded below. Let v be a quasi-subharmonic function on U that is bounded above. If

$$\limsup_{U \ni z \to \zeta} v(z) \le \liminf_{U \ni z \to \zeta} u(z) \,, \quad \text{all} \quad \zeta \in \partial U \,,$$

then $v \le u$ on U.

Proof. Towards obtaining a contradiction, we assume that

$$\alpha = \sup\{v(\phi) - u(\phi) : \phi \in U\}$$

is strictly positive. Since v is bounded above and u below, α is finite. Set

$$E = \{\phi \in U : \limsup_{y \to \phi} [v(y) - u(y)] = \alpha\} \,.$$

Evidently E is a compact subset of U.

Choose $\phi_o \in E$ to be a Shilov boundary point for the algebra A_E. Let N be a compact neighbourhood of ϕ such that $u = \tilde{g}$ on N, where g is the restriction of u to ∂N. The proof of Lemma 5.17 shows that any compact neighbourhood of ϕ_o contained in N has the same property. By shrinking N, then, we can assume that v is quasi-subharmonic on N.

Choose $g \in A$ so that $|g| < 1/2$ on $E \backslash N^o$, while $|g(\phi_1)| > 1$ for some $\phi_1 \in E \cap N^o$. For $c > 0$ sufficiently large, $c(v - u - \alpha) + \log|g| < 0$ on ∂N. Letting $\varepsilon = 1/c$, we obtain

$$v - \alpha + \varepsilon \log|g| \le u \quad \text{on} \quad \partial N \,.$$

If $\phi \in N$, and if σ is a Jensen measure on ∂N for ϕ, then

$$v(\phi) - \alpha + \varepsilon \log|g(\phi)| \le \int [v - \alpha + \varepsilon \log|g|] d\sigma$$

$$\le \int u d\sigma .$$

Taking the infimum over such σ , we obtain

$$v(\phi) - \alpha + \varepsilon \log|g(\phi)| \le u(\phi) .$$

Hence

$$\alpha = \limsup_{\phi \to \phi_1} [v(\phi)-u(\phi)] \le \alpha - \varepsilon \log|g(\phi_1)| .$$

This contradicts $|g(\phi_1)| > 1$. □

According to the definition, a Bremermann function assumes locally the form $\tilde{w}$ for appropriate w . The next theorem shows that often it assumes this form globally.

5.20 Theorem. Let E be a compact subset of M_A such that E^o is disjoint from X . Let u be a log-envelope function on E such that u is a Bremermann function on E^o . Then $u = \tilde{w}$ on E , where $w = u|_{\partial E}$.

Proof. Suppose $c > 0$ and $f \in A$ satisfy $c \log|f| \le u$ on ∂E . Applying Theorem 5.19 to $v = c \log|f|$, we obtain $c \log|f| \le u$ on E^o . Hence $\tilde{w} \le u$. Since u is itself a log-envelope function, we obtain $\tilde{w} = u$. □

5.21 Corollary. Let U be an open subset of $M_A \backslash X$, and let u be a bounded Bremermann function on U . If E is any compact, A-convex subset of U , then $u = \tilde{w}$ on E , where $w = u|_{\partial E}$.

Proof. This follows immediately from Corollary 5.12 and Theorem 5.20. □

5.22 Theorem. The pointwise limit of an increasing net of bounded Bremermann functions is a Bremermann function.

Proof. Suppose that $\{u_\alpha\}$ is the increasing net, converging pointwise to u on U. Let $\phi \in U$, and let N be a compact, A-convex neighbourhood of ϕ. By Corollary 5.21, $u_\alpha = \tilde{w}_\alpha$ on N, where $w_\alpha = u_\alpha|_{\partial N}$. Passing to the limit, we obtain $u = \tilde{w}$ on N, where $w = u|_{\partial N}$. □

The Generalized Dirichlet Problem

Fix an open subset U of $M_A \backslash X$. Let h be a bounded, real-valued function on ∂U. The *generalized Dirichlet problem* is to find a Bremermann function on U that attains the boundary values h on ∂U. We aim to study the solution to the problem given by the classical Perron process.

Define a *subsolution* to be a continuous, locally subharmonic function u on U such that $u^* \leq h$ on ∂U. The upper envelope of all subsolutions will be referred to as the *solution of the generalized Dirichlet problem* with boundary function h, and be denoted by $\check{h}$. Evidently $\check{h}$ is a locally log-envelope function on U. We wish to give conditions on U and on h that guarantee that $\check{h}$ actually attains the boundary values h on ∂U. First we observe that $\check{h}$ is a Bremermann function.

5.23 Lemma. If h is a bounded function on ∂U, then $\check{h}$ is a bounded Bremermann function on U.

Proof. If $h \geq b$ on ∂U, then the constant function b is a subsolution, so that $\check{h} \geq b$. If $h \leq c$ on ∂U, then $w^* \leq c$ on ∂U for any subsolution w. By the maximum principle (Theorem 5.19), $w \le c$ on U. Hence $\check{h} \le c$, and $\check{h}$ is bounded.

Let $p \in U$, and let N be a compact, A-convex neighbourhood of p. Let $v = \check{h}|_{\partial N}$. To show that $\check{h}$ is a Bremermann function, it suffices to show that $\tilde{v} = \check{h}$ on N.

By Corollary 5.12, $\check{h}$ is a log-envelope function on N, so that $\check{h} \le \tilde{v}$. We must obtain the reverse inequality.

Fix $q \in N$, and let $\varepsilon > 0$. Choose $c > 0$ and $f \in A$ such that $c \log|f| < \check{h}$ on ∂N, while $c \log|f(q)| > \tilde{v}(q) - \varepsilon$. For each $y \in \partial N$, there is a continuous, locally subharmonic function u on U such that $u^* \le h$ on ∂U, while $u(y) > c \log|f(y)|$. This estimate persists in a neighbourhood of y. Covering ∂N by a finite number of such neighbourhoods, and taking the maximum of the corresponding functions u, we obtain a continuous, locally subharmonic function w on U such that $w^* \le h$ on ∂U, while $w > c \log|f|$ on ∂N. Define a function w_o on U so that $w_o = w$ on $U \backslash N$, while $w_o = \max(w, c \log|f|)$ on N. Then w_o is continuous and locally subharmonic, and $w_o^* \le h$ on ∂U. Consequently $w_o \le \check{h}$. In particular, $\tilde{v}(q) - \varepsilon < c \log|f(q)| \le \check{h}(q)$. Letting ε tend to zero, we obtain $\tilde{v}(q) \le \check{h}(q)$. □

The theory of barriers can be adapted to give conditions under which the boundary values of $\check{h}$ coincide with h. For simplicity, we assume that U is metrizable, with metric d.

A *barrier* at a point $\zeta \in \partial U$ is a continuous, locally subharmonic function u on U such that

$$\lim_{U \ni z \to \zeta} u(z) = 0 ,$$

while for each $\delta > 0$, the supremum of $u(z)$ for $z \in U$ satisfying $d(z,\zeta) \ge \delta$ is strictly negative. The point

$\zeta \in \partial U$ is a *regular boundary point* of U if there exists a barrier at ζ. From Theorem 1.13, it follows that any $\zeta \in \partial U$ that is a Jensen boundary point for A is a regular boundary point of U.

5.24 Theorem. Suppose that $\zeta \in \partial U$ is a regular boundary point. If h is a bounded function on ∂U, then

$$\limsup_{U \ni p \to \zeta} \check{h}(p) \leq \limsup_{\partial U \ni q \to \zeta} h(q) , \tag{5.16}$$

$$\liminf_{U \ni p \to \zeta} \check{h}(p) \leq \liminf_{\partial U \ni q \to \zeta} h(q) . \tag{5.17}$$

In particular, if each point of ∂U is regular, and if h is continuous on ∂U, then $\check{h}$ assumes the boundary values h continuously at ∂U.

Proof. Let u be a barrier at ζ.

Suppose that $h \geq \alpha$ in a neighbourhood of ζ in ∂U. Then for $c > 0$ large, $v = \alpha + cu$ satisfies $v^* \leq h$ on ∂U. Hence $v \leq \check{h}$ on U, and $\liminf_{z \to \zeta} \check{h}(z) \geq \alpha$. This proves (5.17).

Suppose next that $h \leq \beta$ in a neighbourhood J of ζ in ∂U. Let M be an upper bound for h on ∂U. Then for $c > 0$ large, $M + cu^* \leq \beta$ on $(\partial U)\backslash J$. Let w be any subsolution. Using the estimate $w^* \leq M$ on $(\partial U)\backslash J$, and the estimate $w^* \leq \beta$ on J, we obtain $w^* + cu^* \leq \beta$ on ∂U. By the maximum principle, $w + cu \leq \beta$ on U. Taking the supremum over subsolutions w, we obtain $\check{h} + cu \leq \beta$ on U, and $\limsup_{z \to \zeta} \check{h}(z) \leq \beta$. This proves (5.16). □

In our definition, we have specified that subsolutions be continuous on U. However, we might just as well consider the upper envelope of all locally subharmonic functions w on U such that $w^* \leq h$ on ∂U. This upper envelope might be strictly greater than $\check{h}$. The following theorem gives conditions under which this upper envelope coincides with $\check{h}$.

5.25 Theorem. Let U be an open subset of $M_A \backslash X$ such that every point of ∂U is regular. Suppose that h is a bounded, lower semi-continuous function on ∂U . If w is a locally subharmonic function on U such that $w^* \le h$ on ∂U , then $w \le \check{h}$ on U .

Proof. Define a function u on $\bar{U}$ so that $u = \check{h}$ on U , while $u = h$ on ∂U . By (5.17), u is lower semi-continuous. By Lemma 5.23, u is a Bremermann function on U . Since subharmonic functions are bounded above on compacta, and since w is bounded above near ∂U , w is bounded above on U . From Theorem 5.19 we obtain $w \le u$ on U .

□

In general, the continuity of h does not imply the continuity of $\check{h}$, even if ∂U is regular. Consider for example the algebra of analytic functions on two analytic discs $\{|z| \le 1\}$ and $\{|w| \le 1\}$, attached at their origins $z = w = 0$. For X we take the boundaries of the two discs. The set of Jensen measures for the origin consists of the interval joining the measure $d\theta/2\pi$ on $\{|z| = 1\}$ and $d\psi/2\pi$ on $\{|w| = 1\}$. Each other point has a unique Jensen measure, either a point mass or a Poisson kernel. If the boundary function h is zero on $\{|z| = 1\}$ and 1 on $\{|w| = 1\}$, then $\check{h}$ is zero on $\{|z| \le 1\}$ and 1 on $\{0 < |w| \le 1\}$, so that $\check{h}$ is discontinuous at $z = w = 0$.

Conditions under which $\check{h}$ is continuous are given in the following theorem. We will see in Chapter 6 that the hypotheses are met in $\mathbb{C}^n$.

5.26 Theorem. Let U be an open subset of $M_A \backslash X$. Suppose that if u is any bounded, locally log-envelope function on U, then u^* is locally subharmonic on U. If $h \in C_R(\partial U)$ is such that $\check{h}$ attains the boundary values h continuously at ∂U, then $\check{h}$ is continuous on U.

Proof. By hypothesis, $(\check{h})^*$ is subharmonic, and $(\check{h})^* = h$ on ∂U. By Theorem 5.25, $(\check{h})^* \le \check{h}$ on U. Hence $\check{h}$ is continuous on U. □

References

1. Browder, A. *Introduction to Function Algebras*, W.A. Benjamin, Inc., 1969.
2. Gamelin, T.W. *Uniform algebras on plane sets, in Approximation Theory*, G.G. Lorentz (ed.), Academic Press, New York, 1974, pp.101-149.
3. Gamelin, T.W. Uniform algebras spanned by Hartogs series, *Pacific J. Math.* 62 (1976), 401-417.
4. Gamelin, T.W. and Sibony, N. *Subharmonicity for uniform algebras*, to appear.
5. Wermer, J. Bounded point derivations on certain Banach algebras, *J. Functional Analysis* 1 (1967), 28-36.

6 · Algebras of analytic functions

In this chapter, we aim to study the abstract notion of subharmonicity in the context of algebras generated by analytic functions on compact subsets of $\mathbb{C}^n$. The key to the application of the abstract theory is a theorem of H. Bremermann, asserting that the abstract subharmonic functions essentially coincide with the plurisubharmonic functions. We begin with a review of plurisubharmonic functions.

Plurisubharmonic Functions

Recall that a function u from an open subset D of $\mathbb{C}^n$ to $[-\infty,+\infty)$ is *plurisubharmonic* if u is upper semi-continuous, and the restriction of u to the intersection of D and any complex line is subharmonic. Good references for plurisubharmonic functions are the monographs of L. Hormander [7] and V.S. Vladimirov[9].

The plurisubharmonic functions on D form a convex cone; if u and v are plurisubharmonic and $c > 0$, then $u+v$ and cu are plurisubharmonic. The cone of plurisubharmonic functions includes the cone of functions of the form $c \log |f|$, where $c \geq 0$ and f is analytic on D. The maximum of a finite number of plurisubharmonic functions is plurisubharmonic. The composition of a convex increasing function and a plurisubharmonic function is again plurisubharmonic.

The *complex Hessian matrix* of a smooth function u is the hermitian matrix function

$$H_u(z) = \left(\frac{\partial^2 u}{\partial z_j \partial \bar{z}_k}(z) \right)_{j,k=1}^{n} . \tag{6.1}$$

A smooth function u is plurisubharmonic on D if and only if $H_u(z) \geq 0$ for all $z \in D$. A smooth function u is *strictly plurisubharmonic* on D if the matrix $H_u(z)$ is positive definite at each $z \in D$.

A decreasing limit of plurisubharmonic functions is plurisubharmonic. If u is an arbitrary plurisubharmonic function on D, then there is a sequence of domains $\{D_j\}$ increasing to D, and infinitely differentiable plurisubharmonic functions u_j defined on D_j, such that the sequence $\{u_j\}$ decreases pointwise to u.

Suppose that $\{u_j\}_{j=1}^{\infty}$ is an increasing sequence of plurisubharmonic functions on D, which has a finite pointwise limit u. Since u need not be upper semi-continuous, u is not necessarily plurisubharmonic. However, the upper semi-continuous regularization u^* of u is plurisubharmonic, and $u^* = u$ almost everywhere with respect to volume measure.

Suppose A is a uniform algebra, and that the functions in A are analytic on some open subset D of $\mathbb{C}^n$. Then any function of the form $\max(\log|f_1|,\ldots,\log|f_m|)$, where the f_j's belong to A, is plurisubharmonic on D. Since any function on D that is subharmonic with respect to A can be expressed on compacta as a decreasing limit of such functions, the A-subharmonic functions are plurisubharmonic on D. Later we will consider the problem of determining which plurisubharmonic functions arise from A-subharmonic functions.

Plurisubharmonic Barriers and Pseudoconvex Boundary Points

Let D be a bounded domain in $\mathbb{C}^n$. A *plurisubharmonic barrier* at $\zeta \in \partial D$ is a plurisubharmonic function v on D such that

$$\lim_{D \ni z \to \zeta} v(z) = 0 ,$$

while for each $r > 0$,

$$\sup_{z\in D, |z-\zeta|>r} v(z) < 0 .$$

For example, if ζ is a peak point for an analytic function f on D, then $|f|-1$ forms a plurisubharmonic barrier at ζ.

Note that the notion of plurisubharmonic barrier is local. If a plurisubharmonic function v, defined in the part of D near ζ, has the properties above, then for $\delta > 0$ sufficiently small, the function defined to be $\max(v,-\delta)$ near ζ and $-\delta$ away from ζ is a plurisubharmonic barrier at ζ.

A point $\zeta \in \partial D$ is a *smooth point* of ∂D if there is an open ball B containing ζ and a smooth function ρ on B such that $d\rho \neq 0$, while $B \cap D = \{\rho < 0\}$. The point ζ is a *strictly pseudoconvex boundary point* of D if the complex Hessian matrix associated with ρ as in (6.1) determines a positive definite bilinear form on the complex tangent space to ∂D at ζ, that is if $\sum \frac{\partial^2\rho}{\partial z_j \partial \bar{z}_k}(\zeta)\lambda_j\bar{\lambda}_k > 0$ for all nonzero $\lambda \in \mathbb{C}^n$ that satisfy $\sum \frac{\partial\rho}{\partial z_j}(\zeta)\bar{\lambda}_j = 0$. This definition depends neither on the smooth defining function ρ nor on the analytic system of coordinates at ζ. Replacing ρ by $e^{t\rho}-1$ for t large, we can always arrange that the complex Hessian matrix of a defining function at a strictly pseudoconvex boundary point is positive definite.

6.1 Lemma. If ζ is a smooth, strictly pseudoconvex boundary point of D, then there is a plurisubharmonic barrier at ζ.

Proof. Choose B and ρ as above, so that the complex Hessian matrix of ρ is positive definite on B. Expanding ρ in a Taylor series, we find an analytic quadratic polynomial f such that

$$\rho(z) = \mathrm{Re}(f(z)) + \sum \frac{\partial^2\rho}{\partial z_j \partial \bar{z}_k}(\zeta)(z_j-\zeta_j)(\bar{z}_k-\bar{\zeta}_k) + o(|z-\zeta|^2) .$$

Since the Hessian of ρ is positive definite at ζ, there exists $c > 0$ such that

$$\mathrm{Re}(f(z)) \leq \rho(z) - c|z-\zeta|^2 \tag{6.2}$$

for z near ζ. It follows that $\mathrm{Re}(f)$ is a barrier at ζ, at least locally, and hence there is a global barrier at ζ. □

There is a converse to Lemma 6.1.

6.2 Lemma. Let ζ be a smooth boundary point of D. Suppose there is a smooth strictly plurisubharmonic function v on a ball B containing ζ such that $v \leq 0$ on $B \cap D$, and $v(\zeta) = 0$. Then ζ is a strictly pseudoconvex boundary point of D.

Proof. We follow H.Rossi[8]. Making an analytic affine change of variables, we can assume that $\zeta = 0$, and that D is defined by a function ρ of the form

$$\rho(x_1 + iy_1, \ldots, x_n + iy_n) = x_1 + o(|z|) .$$

We must show that $\Sigma \frac{\partial \rho}{\partial z_j \partial \bar{z}_k}(0)\lambda_j\bar{\lambda}_k > 0$ for nonzero $\lambda \in \mathbb{C}^n$ satisfying $\lambda_1 = 0$. Making a linear analytic change of the variables $z_2, \ldots, z_n$, we can arrange also that $\lambda_3 = \ldots \lambda_n = 0$. It suffices then to show that

$$\frac{\partial^2 \rho}{\partial z_2 \partial \bar{z}_2}(0) = \frac{1}{4}\left[\frac{\partial^2 \rho}{\partial x_2^2}(0) + \frac{\partial^2 \rho}{\partial y_2^2}(0)\right] > 0 . \tag{6.3}$$

Near 0, ∂D forms a manifold, which is coordinatized by the variables $(y_1, x_2, y_2, \ldots, x_n, y_n)$. Let $X_1 = X_1(y_1, x_2, \ldots, y_n)$ by the x_1-coordinate on ∂D, regarded as a function of the remaining variables, so that

$$\rho(X_1(y_1,x_2,\ldots,y_n),y_1,x_2,\ldots,y_n) = 0 \ .$$

For a function ψ near 0, denote by ψ^r the restriction of ψ to ∂D, as a function of the coordinates $y_1,x_2,\ldots,y_n$, so that

$$\psi^r(y_1,x_2,\ldots,y_n) = \psi(X_1(y_1,x_2,\ldots,y_n),y_1,x_2,\ldots,y_n) \ .$$

From the chain rule we obtain

$$\frac{\partial\psi^r}{\partial x_2} = \frac{\partial\psi}{\partial x_2} + \frac{\partial\psi}{\partial x_1}\frac{\partial X_1}{\partial x_2} ,$$

$$\frac{\partial^2\psi^r}{\partial x_2^2} = \frac{\partial^2\psi}{\partial x_2^2} + \frac{\partial^2\psi}{\partial x_1\partial x_2}\frac{\partial X_1}{\partial x_2} + \frac{\partial\psi}{\partial x_1}\frac{\partial^2 X_1}{\partial x_2^2} + \frac{\partial^2\psi}{\partial x_2^2}\frac{\partial X_1}{\partial x_2} + \frac{\partial^2\psi}{\partial x_1^2}\left(\frac{\partial X_1}{\partial x_2}\right)^2 .$$

Evaluating at 0, and noting that $\frac{\partial X_1}{\partial x_2}(0) = 0$, we obtain

$$\frac{\partial^2\psi^r}{\partial x_2^2}(0) = \frac{\partial^2\psi}{\partial x_2^2}(0) + \frac{\partial\psi}{\partial x_1}(0)\frac{\partial^2 X_1}{\partial x_2^2}(0) \ . \tag{6.4}$$

Applying (6.4) to ρ, and noting that $\rho^r = 0$ while $\frac{\partial\rho}{\partial x_1}(0) = 1$, we obtain

$$0 = \frac{\partial^2\rho}{\partial x_2^2}(0) + \frac{\partial^2 X_1}{\partial x_2^2}(0) \ . \tag{6.5}$$

Applying (6.4) to the plurisubharmonic function u, and now using (6.5), we obtain

$$\frac{\partial^2 u^r}{\partial x_2^2}(0) = \frac{\partial^2 u}{\partial x_2^2}(0) - \frac{\partial u}{\partial x_1}(0)\frac{\partial^2\rho}{\partial x_2^2}(0) \ .$$

Since u attains its maximum over ∂D at 0, at least locally, we have $\frac{\partial^2 u^r}{\partial x_2^2}(0) \le 0$, and

$$\frac{\partial^2 u}{\partial x_2^2}(0) \le \frac{\partial u}{\partial x_1}(0) \frac{\partial^2 \rho}{\partial x_2^2}(0) .$$

There is a similar estimate for $\frac{\partial^2 u}{\partial y_2^2}(0)$. Adding the estimates, we obtain

$$\frac{\partial^2 u}{\partial x_2^2}(0) + \frac{\partial^2}{\partial y_2^2}(0) \le \frac{\partial u}{\partial x_1}(0) \left[\frac{\partial^2 \rho}{\partial x_2^2}(0) + \frac{\partial^2 \rho}{\partial y_2^2}(0)\right] .$$

The term on the left is $4 \frac{\partial^2 u}{\partial z_2 \partial \bar{z}_2}$, which is strictly positive on account of the strict plurisubharmonicity of u . Since $\frac{\partial u}{\partial x_1}(0) \ge 0$, we obtain (6.3), as required.

□

S_δ-sets

A compact subset K of $\mathbb{C}^n$ is an *S_δ-set* if K is the limit of a decreasing sequence of open sets, each component of which is a domain of holomorphy. We denote by $\mathcal{O}(K)$ the algebra of functions that are analytic in a neighbourhood of K , and by $H(K)$ the uniform closure of $\mathcal{O}(K)$ in $C(K)$. It is the algebra $H(K)$ that we are interested in.

Any compact subset of the complex plane is an S_δ-set, and in this case $H(K)$ coincides with the usual algebra $R(K)$. Any compact polynomially convex subset of $\mathbb{C}^n$ is an S_δ-set, and in this case the algebra $H(K)$ coincides with the algebra $P(K)$ generated by the polynomials in $z_1,\ldots,z_n$.

6.3 Theorem. If K is an S_δ-set, then the maximal ideal space of $H(K)$ coincides with K .

Proof. This follows readily from the fact that every nonzero complex-valued homomorphism of the algebra of analytic functions on a domain of holomorphy is the evaluation homomorphism at some point of the domain.

□

Now fix a compact subset K of $\mathbb{C}^n$. Let ζ be a smooth boundary point of K, so that there is an open ball B containing ζ and a smooth function ρ on B with $d\rho \neq 0$ and $B \cap K = \{\rho = 0\}$. If the complex Hessian $H_\rho(\zeta)$ is not positive (indefinite) on the complex tangent space to ∂K at ζ, then there is a ball B' containing ζ such that all functions satisfying the tangential Cauchy-Riemann equation on $\{\rho = 0\}$ extend to be analytic in $B' \cap \{\rho > 0\}$ (cf. [6, Theorem 2.6.13]). In particular, all functions in $\mathcal{O}(K)$ extend holomorphically to B', so that K cannot be the spectrum of $H(K)$. We conclude that if K is an S_δ-set, then the Hessian of any smooth defining function is positive on the complex tangent space at any smooth point of ∂K, so the smooth points of ∂K are pseudoconvex.

6.4 Theorem (Rossi's Theorem). Let K be an S_δ-subset of $\mathbb{C}^n$, and let N be the relatively open subset of ∂K consisting of the smooth boundary points. Then the intersection of N and the Shilov boundary of $H(K)$ coincides with the adherence in N of the strictly pseudoconvex boundary points.

Proof. Let ζ be a smooth strictly pseudoconvex boundary point of K. There is then a smooth defining function ρ for K near ζ such that $H_\rho(\zeta)$ is positive definite. As earlier, there is a quadratic analytic polynomial f such that (6.2) is valid. Then $\exp(f)$ is a local peaking function at ζ. By Rossi's local peak point theorem, ζ is a peak point for $H(K)$, and in fact there is a function peaking at ζ in $\mathcal{O}(K)$ (cf. [5]). In particular, ζ belongs to the Shilov boundary of $H(K)$.

Conversely, suppose that $\zeta \in N$ belongs to the Shilov boundary of $H(K)$. Let $\varepsilon > 0$ be small, so that ∂K has a defining function on the open ball B_ε centered at ζ with radius ε. We must find a strictly pseudoconvex boundary point of K within B_ε .

Choose $f \in \mathcal{O}(K)$ such that $\|f\| = 1$, while $|f|$ is small on $K \backslash B_\varepsilon$. For $\delta > 0$ sufficiently small, the function

$$u(z) = \delta|z|^2 + \log|f(z)|$$

attains its maximum over K at some point $\xi_1 \in B_\varepsilon \cap \partial K$. We can assume that $f(\xi_1) \neq 0$, so that u is smooth and strictly plurisubharmonic in a neighbourhood of ξ_1 . By Lemma 6.2, ∂K is strictly pseudoconvex at ξ_1 .

□

There is a geometric argument which makes it plausible that the Shilov boundary should fail to meet an open set on ∂K on which the complex Hessian is degenerate. It turns out that "most" of such points lie on analytic varieties in ∂K . More precisely, if Q is a relatively open subset of ∂K consisting of smooth boundary points, such that the complex Hessian $H_\rho(\zeta)$ for a defining function ρ has constant rank r on the complex tangent spaces T_ζ at $\zeta \in Q$, then each point of Q has a neighbourhood that is fibered by analytic varieties of dimension r . Indeed, let M_ζ be the space of vectors in T_ζ annihilated by $H_\rho(\zeta)$, so that the M_ζ are complex subspaces of T_ζ of dimension r . One checks that the T_ζ's form an involutive distribution. Frobenius' Theorem then shows that through each $\zeta \in Q$, there passes a manifold M in Q such that the tangent space to M at each point $\zeta \in M$ coincides with M_ζ . Since the M_ζ's are complex subspaces, M is a complex analytic manifold. For details, see [4].

Bremermann's Theorem

Our next task is to prove a theorem of Bremermann

relating plurisubharmonicity and $H(K)$-subharmonicity. The proof we will give is, in broad outline, the same as that of Bremermann[2]. We begin by establishing several lemmas.

6.5 Lemma. Let U be a domain of holomorphy in $\mathbb{C}^n$, and let v be plurisubharmonic on U. Then each component of the set $\{v < 0\}$ is a domain of holomorphy.

Proof. Since U is pseudoconvex, there exists a plurisubharmonic function ψ on U that tends to $+\infty$ at ∂U. Let χ be a convex increasing function from $(-\infty, 0]$ to $(-\infty, +\infty]$ such that $\chi(0) = +\infty$. Then $\chi \circ v$ is a plurisubharmonic function on $\{v < 0\}$, and $\psi + \chi \circ v$ tends to $+\infty$ at the boundary of $\{v < 0\}$. Since $\{v < 0\}$ has a plurisubharmonic exhaustion function, each component is a domain of holomorphy. □

6.6 Lemma. Let D be a domain of holomorphy in $\mathbb{C}^n$, and let u be plurisubharmonic on D. Define

$$D^* = \{(z, \zeta) \in \mathbb{C}^{n+1} : z \in D, |\zeta| < e^{-u(z)}\} . \tag{6.6}$$

Then D^* is a domain of homomorphy.

Proof. Apply the preceding lemma to the plurisubharmonic function $v(z, \zeta) = u(z) + \log|\zeta|$ on $D \times \mathbb{C}$.

□

6.7 Lemma. Let D be a domain of holomorphy in $\mathbb{C}^n$, let u be plurisubharmonic on D, and define D^* as in (6.6). Then there is an analytic function

$$F(z, \zeta) = \sum_{j=0}^{\infty} f_j(z) \zeta^j , \quad z \in D, |\zeta| < e^{-u(z)} , \tag{6.7}$$

on D^* such that for each fixed $z \in D$, the radius of convergence $R_F(z)$ of the series (6.7) is given by

$$R_F(z) = e^{-u(z)} , \quad z \in D . \tag{6.8}$$

Proof. Let $\{K_m\}$ be a sequence of compact holomorphically convex subsets of D^* that increases to D^*, and let E_m be the projection of K_m into D. For fixed $z_0 \in E_m$, choose $(z_0,\zeta_0) \in D^*\backslash K_m$ such that $\zeta_0 \neq 0$. Choose f holomorphic on D^* such that $\zeta_0 f(z_0,\zeta_0) = 1$, while on K_m $|\zeta f(z,\zeta)| \leq 1/2$. Since ζf is not identically 1 on the slice $\{z = z_0\}$, the variety $\{\zeta f = 1\}$ projects onto a neighbourhood of z_0. Covering E_m by a finite collection of such neighbourhoods, we obtain a finite collection $f_1,\ldots,f_k$ of analytic functions on D^* such that $|\zeta f_j| < 1/2$ on K_m, $1 \leq j \leq k$, while for each $z_0 \in E_m$, there exists $(z_0,\zeta_0) \in D^*$ and an index j such that $\zeta_0 f_j(z_0,\zeta_0) = 1$. Define

$$F_m = \prod_{j=1}^{k} [1 - (\zeta f_j)^N] ,$$

where the integer N is chosen so large that $|1-F_m| < 1/2^m$ on K_m. Then for each $z_0 \in E_m$, the function $F_m(z_0,\zeta)$ has a zero in the disc $\{|\zeta| < e^{-u(z)}\}$, while $F_m(z_0,0) = 1$. The product

$$F(z,\zeta) = \prod_{m=1}^{\infty} F_m(z,\zeta)$$

then converges normally on D^*. For each $z_0 \in D$, the function $F(z_0,\zeta)$ has an infinite sequence of zeros in the disc $\{|\zeta| < e^{-u(z)}\}$, while $F(z_0,0) = 1$. It follows that the radius of convergence of the Hartogs series expansion (6.7) of F is given by (6.8).

□

6.8 Theorem (Bremermann's Theorem). Let K be an S_δ-subset of $\mathbb{C}^n$, and let u be a function that is defined, continuous and plurisubharmonic in a neighbourhood of K. Then u is subharmonic with respect to $H(K)$, that is, u can be approximated uniformly on K by functions of the form $\max(c_1 \log|f_1|,\ldots,c_m \log|f_m|)$, where $c_1,\ldots,c_m > 0$ and $f_1,\ldots,f_m$ are analytic in a neighbourhood of K.

Proof. Choose a domain of holomorphy $D \supset K$ such that u is continuous and plurisubharmonic on D. Define D^* as in (6.6) and choose a holomorphic function F on D^* as in Lemma 6.7. Choose $r > 0$ and $M > 0$ such that F is defined on the set $\{(z,\zeta) : z \in K, |\zeta| \le r\}$ and bounded in modulus by M there. The Cauchy estimates for the coefficients of the power series (6.7) then become

$$|f_j(z)|r^j \le M, \quad z \in K, j \ge 0 .$$

It follows that the functions

$$u_j(z) = \frac{1}{j} \log|f_j(z)| , \quad z \in K ,$$

are bounded above on K, uniformly in j. Since

$$u(z) = -\log R_F(z) = \limsup_{j\to\infty} u_j(z) ,$$

we conclude from Corollary 1.7 that u is $H(K)$-subharmonic.

□

6.9 Corollary. Let A be a uniform algebra. Let U be an open subset of M_A that is homeomorphic to an open subset of $\mathbb{C}^n$ in such a way that the functions in A are analytic on U and give local coordinates at each point of U. Then the functions on U that are locally subharmonic with respect to A are the plurisubharmonic functions on U .

Proof. We have already observed that the locally A-subharmonic functions on D are plurisubharmonic on D .

Let $z \in D$. Since A gives local coordinates at z , there is a compact neighbourhood of z on which each of the coordinate functions $z_1, \ldots, z_n$ is approximable uniformly by functions in A . We can take the neighbourhood to be a closed ball B centered at z . If u is plurisubharmonic on D , then by Bremermann's Theorem, u is H(B)-subharmonic, and since A is dense in H(B) , u is A-subharmonic on B . Hence plurisubharmonic functions are locally A-subharmonic. □

Note that we have used Bremermann's Theorem only for closed balls. It would be of interest to find an elementary proof of Bremermann's Theorem for this case, which does not depend on the characterization of domains of holomorphy in terms of pseudoconvexity.

Subharmonicity with Respect to H(K)

Fix a compact S_δ-subset K of $\mathbb{C}^n$. We wish to combine Bremermann's Theorem with the localization theorem of Chapter 5, to obtain conditions under which a given function is H(K)-subharmonic, that is, subharmonic with respect to the algebra H(K) . We begin with the following.

6.10 Theorem. Let K be an S_δ-subset of $\mathbb{C}^n$. Then the functions on K^o that are H(K)-subharmonic are precisely the plurisubharmonic functions on K^o .

Proof. In view of Corollary 6.9, it suffices to show that a locally H(K)-subharmonic function on K^o is H(K)-subharmonic. This follows immediately from Corollary 5.11, once we show that K^o is H(K)-convex.

Let E be a compact subset of K^o. Let $\{D_k\}$ be a sequence of domains of holomorphy that decreases to K, and let E_k by the $\mathcal{O}(D_k)$-convex hull of E. Then the distance from E_k to ∂D_k coincides with the distance from E to ∂D_k, by one of the characterizing properties of domains of holomorphy. Since $\hat{E} \subset E_k$, we find in the limit that the distance from $\hat{E}$ to ∂K coincides with the distance from E to ∂K, and $\hat{E}$ is a compact subset of K^o.

□

6.11 Theorem. Let K be an S_δ-subset of $\mathbb{C}^n$. If every point of ∂K is a Jensen boundary point for $H(K)$, then the H(K)-subharmonic functions are precisely the upper semi-continuous functions from K to $[-\infty,+\infty)$ that are plurisubharmonic on K^o.

Proof. This follows immediately from Theorem 6.10 and Theorem 5.14. □

Now suppose that ζ is a boundary point of K, and that u is an upper semi-continuous function defined on a neighbourhood of ζ in K. Under what conditions is u a locally H(K)-subharmonic function near ζ? The most obvious condition is that u extend to be plurisubharmonic in a ball B containing ζ. Then u is H(B)-subharmonic, and hence $H(B \cap K)$-subharmonic.

Another condition is that ζ be a smooth point of ∂K, that u be continuous near ζ, and that u be plurisubharmonic on the part of K^o near ζ. Suppose these circumstances hold, let $\varepsilon > 0$ be small, and let ν be the outer unit normal vector to K at ζ. Then $u_\varepsilon(z) = u(z-\varepsilon\nu)$ is plurisubharmonic in a neighbourhood of $B \cap K$ for some fixed small ball B centered at ζ. Consequently

u_ε is locally H(K)-subharmonic on $B \cap K$. Since u_ε converges uniformly to u, also u is locally H(K)-subharmonic on $B \cap K$, by Theorem 5.14. This leads us to the following theorem.

6.12 Theorem. Let K be a compact S_δ-subset of $\mathbb{C}^n$ such that ∂K is smooth. Then the continuous H(K)-subharmonic functions on K are precisely the continuous functions on K that are plurisubharmonic on the interior of K.

Proof. The previous considerations show that any continuous function on K that is plurisubharmonic on K^o is locally H(K)-subharmonic. By Theorem 5.9, such a function is H(K)-subharmonic. The reverse implication is clear.

□

Now let K be a compact S_δ-set with smooth boundary. It has already been seen that the strictly pseudoconvex boundary points of K are peak points for $H(K)$. There may also be peak points for $H(K)$ which are not strictly pseudoconvex boundary points. Consider for instance the ball with the bulge, $\{|z_1|^2 + |z_2|^4 \le 1\}$, in $\mathbb{C}^2$. The function $|z_1|^2 + |z_2|^4 - 1$ is a defining function for the boundary, and its complex Hessian is diagonal, with entries 1 and $4|z_2|^2$. On the circle $\{|z_1| = 1, z_2 = 0\}$, the complex tangent space is spanned by $\frac{\partial}{\partial z_2}$, so that the Hessian matrix annihilates the complex tangent space, and points on the circle are not strictly pseudoconvex boundary points. However, each point of the circle is a peak point for $H(K)$, and in fact $(1+\bar{\lambda} z_1)/2$ peaks at $(\lambda,0)$, where $|\lambda| = 1$.

One problem that is currently unsolved is the following. Suppose that ∂K is smooth, and that each point of ∂K, with one exception, is a strictly pseudoconvex boundary point. Is the exceptional point necessarily a peak point for $H(K)$? It turns out that the exceptional point *is* necessarily a Jensen boundary point for $H(K)$. In fact, much more is true.

6.13 Theorem. Let K be a compact S_δ-subset of $\mathbb{C}^n$ with smooth boundary, and let T be the set of points on ∂K that are not strictly pseudoconvex boundary points. Then every Jensen measure for a point of T is supported by T . In particular, any isolated point of T is a Jensen boundary point for $H(K)$.

Proof. The second assertion follows from the first and the localization theorem for the Jensen boundary. (See the remarks after Theorem 5.15.)

Let σ be a Jensen measure for $\zeta \in T$, and let ξ be a strictly pseudoconvex boundary point for K . By perturbing a strictly pseudoconvex defining function for ∂K near ξ , we may obtain a compact set J with smooth boundary, and an open ball B centered at ξ , such that $J \supseteq K$, $J \backslash B = K \backslash B$, $\xi \in J^o$, and each point of $B \cap \partial J$ is a strictly pseudoconvex boundary point of J . In these circumstances, K.Diederich and J.E.Fornaess[3] have constructed a continuous plurisubharmonic exhaustion function for J , that is, a continuous function ρ on J such that $\rho = 0$ on ∂J , $\rho < 0$ on J^o , and ρ is plurisubharmonic on J^o . By Theorem 6.12, ρ is $H(K)$-subharmonic. Hence $0 = \rho(\zeta) \leq \int \rho d\sigma$, and σ is evidently supported on the set $\{\rho = 0\} \cap K$. In particular, the closed support of σ is disjoint from K^o , and moreover ξ does not belong to the closed support of σ . Since $\xi \in (\partial K) \backslash T$ is arbitrary, σ is supported on T .

□

Bremermann Functions

Let D be an open subset of $\mathbb{C}^n$. A lower semi-continuous function u from D to $(-\infty,+\infty]$ is a *Bremermann function on* D if each point of D is included in an open ball B , $\bar{B} \subset D$, such that $u|_B$ is the upper envelope of all functions of the form $c \log|f|$, where $c > 0$, f is an analytic polynomial, and $c \log|f| < u$ on ∂B . If D is bounded, this occurs if an only if u is a Bremermann function with respect to the uniform algebra spanned by the analytic polynomials on any ball or polydisc containing D .

The *locally log-envelope functions on* D are defined similarly, to be the lower semi-continuous functions from D to $(-\infty,+\infty]$ that are locally the upper envelopes of functions of the form $c \log|f|$, where $c > 0$ and f is an analytic polynomial. As in the proof of Corollary 6.9, one sees that if A is any uniform algebra that contains D as an open subset of its maximal ideal space, such that the functions in A are analytic on D and give local coordinates, then the locally log-envelope functions on D (as just defined) coincide with the locally log-envelope functions on D with respect to A , and the Bremermann functions on D (as defined above) coincide with the Bremermann functions for A on D . In particular, the class of locally log-envelope functions and the class of Bremermann functions are invariant under analytic changes of variables.

Return now to the definition of Bremermann function. There are various ways to restate the condition on $u|_B$. The condition is simply that $u|_B$ coincides with the solution $\tilde{v}$ of the A-Dirichlet problem with boundary data $v = u|_{\partial B}$, where A is the algebra of analytic polynomials. From the maximum principle (Theorem 5.19), we see that $u|_B$ is also the upper envelope of all plurisubharmonic functions w on B such that $w^* \leq u$ on ∂B .

The condition on $u|_B$ can also be restated in terms of polynomial hulls. Let

$$Y = \{(z,\zeta) \in \mathbb{C}^{n+1} : z \in \partial B,\ |\zeta| \le e^{-u(z)}\} . \tag{6.9}$$

Then the condition $u = \tilde{v}$ on B, where $v = u|_{\partial B}$, holds if and only if the polynomial convex hull of Y is given by

$$\{(z,\zeta) \in \mathbb{C}^{n+1} : z \in \bar{B},\ |\zeta| \le e^{-u(z)}\} . \tag{6.10}$$

This follows immediately from Theorem 5.4, since the polynomial convex hull of Y coincides with the maximal ideal space of the uniform algebra on Y generated by the polynomials.

In the case $n = 1$, the condition $u = \tilde{v}$, where $v = u|_{\partial B}$, means simply that $u|_B$ is the Poisson integral of its boundary values on ∂B. It follows that the Bremermann functions on a domain in the complex plane are simply the harmonic functions.

In particular, the Bremermann functions of one complex variable are real analytic. We will show presently that this result fails miserably for Bremermann functions of several complex variables. Nevertheless, the following characterization of the smooth Bremermann functions is available.

6.14 Theorem. A smooth function u on D is a Bremermann function if and only if u is plurisubharmonic, and the complex Hessian matrix of u is singular at each point of D.

Proof. Let B be an open ball with closure included in D, and let K be the set defined by (6.7). The piece of ∂K lying over B is smooth, with defining function

$$\rho(z,\zeta) = u(z) + \log|\zeta| .$$

The complex Hessian matrix H_ρ of ρ can be expressed simply in terms of that of u :

$$H_\rho(z,\zeta) = \left(\begin{array}{c|c} H_u(z) & 0 \\ \hline 0 & 0 \end{array}\right).$$

Recall that the complex tangent space $T_{(z,\zeta)}$ at $(z,\zeta) \in \partial K$ consists of vectors in $\mathbb{C}^{n+1}$ orthogonal to $d\rho$. Since $d\rho$ has a nonzero component in the ζ-direction, it is easy to see that the rank of $H_\rho(z,\zeta)$ on $T_{(z,\zeta)}$ coincides with the rank of $H_u(z)$. In particular, (z,ζ) is a strictly pseudoconvex point of ∂K if and only if $H_u(z)$ is positive definite.

Now suppose that u is a smooth Bremermann function. From the definition, u is plurisubharmonic, so that $H_u(z) \geq 0$ on B . Moreover, the set K defined by (6.10) is the polynomial hull of the set Y defined by (6.9). Since the Shilov boundary of $P(K)$ is included in Y , no point of ∂K lying over B can be a strictly pseudoconvex point of ∂K , by Rossi's Theorem. (This assertion is also easy to establish directly.) We conclude that $H_u(z)$ is singular at each $z \in B$.

Conversely, suppose that u is a smooth plurisubharmonic function, and that $H_u(z)$ is singular at each point of B . Since u is plurisubharmonic, u is subharmonic with respect to the algebra on $\bar{B}$ generated by the analytic polynomials. By Theorem 5.4, K coincides with the maximal ideal space of the algebra generated by the analytic polynomials on K . Now H_u is singular, so that also $H_\rho(z,\zeta)$ is singular at each point $(z,\zeta) \in \partial K$ lying over B . By Rossi's Theorem, the Shilov boundary of $P(K)$ is included in the set Y defined by (6.9). It follows that K coincides with the polynomial hull of Y , and u is a

Bremermann function. □

The following theorem provides us with a fairly wide class of Bremermann functions.

6.15 Theorem. Let D be a domain in $\mathbb{C}^n$, let $1 \le k < n$, and let F be an analytic map of D into a domain D' in $\mathbb{C}^k$. If w is any locally bounded, locally log-envelope function on D', then $w \circ F$ is a Bremermann function on D.

Proof. Suppose first that w is a smooth plurisubharmonic function. Then $w \circ F$ is also smooth and plurisubharmonic, and the rank of the complex Hessian matrix of $w \circ F$ is at most k at each point of D. By Theorem 6.14, $w \circ F$ is a Bremermann function.

For the general case, one expresses w locally as an increasing limit of smooth plurisubharmonic functions w_α. Each $w_\alpha \circ F$ is a Bremermann function, and by Theorem 5.22, the increasing limit of Bremermann functions is a Bremermann function, so that $w \circ F$ is a Bremermann function.

□

Now let χ be a convex function of a real variable. Define u on $\mathbb{C}^n$ by

$$u(z_1, \ldots, z_n) = \chi(x_1) , \quad z_1 = x_1 + iy_1 .$$

Regarded as a function of the complex variable z_1, χ is subharmonic. By Theorem 6.15, or by a simple direct argument, u is a Bremermann function on $\mathbb{C}^n$ for $n \ge 2$. By choosing χ to be continuous but not smooth, we obtain in this manner a continuous Bremermann function that is not smooth. In like manner, we can find a Bremermann function with any specified degree of smoothness, which fails to have a higher degree of smoothness.

Nothing could be easier than producing a discontinuous Bremermann function. Define w on $\mathbb{C}$ so that $w(0) = 0$, while $w = 1$ on $\mathbb{C}\backslash\{0\}$. Near 0, w is the increasing limit of the subharmonic functions $|z|^\alpha$, as α decreases to 0. Hence $u(z_1,\ldots,z_n) = w(z_1)$ is a discontinuous Bremermann function on $\mathbb{C}^n$, $n \ge 2$.

One may construct more ill-behaved examples, for which the upper semi-continuous regularization is not continuous, as follows. Let $a_j \downarrow 0$, let $\varepsilon_j \downarrow 0$ rapidly, let $M > 0$ be large, define the subharmonic function $v(z) = \Sigma\varepsilon_j \log |z-a_j|$, and set $w = \max(v,-M)$ on $\mathbb{C}\backslash\{0\}$, $w(0) = -M$. Then w is lower semi-continuous, and w is easily seen to be quasi-subharmonic, so that w is a locally log-envelope function. For $n \ge 2$, the function $u(z_1,\ldots,z_n) = w(z_1)$ is a Bremermann function that is discontinuous at $z_1 = 0$, and u^* is also discontinuous at $z_1 = 0$.

Bremermann's Generalized Dirichlet Problem

Let D be a bounded domain in $\mathbb{C}^n$, and let h be a bounded, real-valued function on ∂D. The solution to the generalized Dirichlet problem defined in Chapter 5 is the upper envelope $\check{h}$ of the continuous plurisubharmonic functions u on D satisfying $u^* \le h$ on ∂D. It is natural to consider also the upper envelope $\tilde{h}$ of *all* plurisubharmonic functions u on D satisfying $u^* \le h$ on ∂D. The upper semi-continuous regularization of $\tilde{h}$ is a plurisubharmonic function on D that is *Bremermann's solution to the generalized Dirichlet problem* with boundary data h.

Evidently $\check{h} \le \tilde{h}$. Often the functions $\check{h}$ and $\tilde{h}$ coincide. For instance, it follows from Theorem 5.25 that $\check{h} = \tilde{h}$ whenever h is lower semi-continuous and there is a continuous plurisubharmonic barrier at each point of ∂D.

Another condition guaranteeing that $\check{h} = \tilde{h}$ is given in the following theorem of J.B.Walsh[10]. The abstract analogue of Walsh's theorem, Theorem 5.26, does not quite suffice to yield the Walsh theorem in the case at hand.

6.16 Theorem. Let $h \in C_R(\partial D)$. Suppose that for each $\zeta \in \partial D$, $\tilde{h}(z)$ tends to $h(\zeta)$ as $z \in D$ tends to ζ . Then $\tilde{h}$ is continuous on D . In particular, $\tilde{h} = \check{h}$.

Proof. Let $\varepsilon > 0$. Choose $\delta > 0$ so that if $z,w \in D$ lie in a (2δ)-neighbourhood of ∂D , and if $|z-w| < \delta$, then $|\tilde{h}(z)-\tilde{h}(w)| < \varepsilon$. Fix $y \in \mathbb{C}^n$ such that $|y| < \delta$, and define v_y on D so that $v_y = \tilde{h}$ on a δ-neighbourhood of ∂D , while

$$v_y(z) = \max\{\tilde{h}(z), \tilde{h}(z+y) - \varepsilon\}$$

elsewhere. Since the term $\tilde{h}(z)$ dominates when z lies on the boundary of the δ-neighbourhood of ∂D , we see that v_y is plurisubharmonic. Since $v_y = \tilde{h}$ near ∂D , while $\tilde{h}$ tends to h at ∂D , the definition of $\tilde{h}$ shows that $v_y \le \tilde{h}$ on D . Consequently $\tilde{h}(w) - \varepsilon \le \tilde{h}(z)$ whenever $z,w \in D$ satisfy $|z-w| < \delta$. It follows that $\tilde{h}$ is continuous on D . □

If we specialize to S_δ-sets with regular boundaries, we obtain the following.

6.17 Theorem. Let K be a compact S_δ-subset of $\mathbb{C}^n$ such that every point of ∂K is a Jensen boundary point for $H(K)$. Let u be a bounded, lower semi-continuous function on ∂K . Then $\tilde{u} = \check{u} = \tilde{u}$ on K^o . Furthermore, if u is continuous on ∂K , then $\tilde{u}$ is continuous on K .

Proof. From the definitions, we have $\tilde{u} \le \check{u} \le \tilde{u}$ on K^o .

Let v be any plurisubharmonic function on K^o such that $v^* \le u$ on ∂K^o. Since every point of ∂K is a Jensen boundary point, $u = \tilde{u}$ on ∂K. Hence $v^* \le \tilde{u}$ on ∂K^o, and furthermore $\tilde{u}$ is a Bremermann function on K^o, by Lemma 5.17. By the maximum principle (Theorem 5.19), $v \le \tilde{u}$ on K^o. Passing to the upper envelope, we obtain $\tilde{u} \le \tilde{u}$ on K^o. This proves the first assertion of the theorem. The second assertion now follows from Theorem 6.16, or from Theorem 5.26. □

Now let us focus on the open unit ball B in $\mathbb{C}^n$, for $n \ge 2$. The existence of discontinuous Bremermann functions shows that $\tilde{u}$ need not be continuous on B when u is discontinuous on ∂B. It turns out, though, that if u is continuous on ∂B, then $\tilde{u}$ has certain regularity properties on B. E.Bedford and B.A.Taylor[1] have shown in this case that the first partial derivatives of $\tilde{u}$ exist and are Lipschitz continuous on compact subsets of B. In particular, the second partial derivatives of $\tilde{u}$ exist almost everywhere with respect to volume measure.

On the other hand, the second partial derivatives of $\tilde{u}$ need not exist everywhere on B, even though u is a real-analytic function on ∂B. To see this, consider first the function w of one complex variable, defined by

$$w(re^{i\theta}) = \begin{cases} (4r^2-1)^2, & r \ge 1/2, \\ 0, & r \le 1/2. \end{cases}$$

Then w is continuous. Since $w \ge 0$, while $\Delta w = 32[8r^2-1] > 0$ for $r > 1/2$, w is seen to be subharmonic. Define

$$u(z) = (4z_1\bar{z}_1 - 1)^2, \qquad z \in \partial B.$$

Since $u \ge 0$, also $\tilde{u} \ge 0$. The maximum principle, applied for fixed values of z_1, shows that

$$\tilde{u}(z) \le (4|z_1|^2 - 1)^2 , \quad z \in B .$$

Since $u(z) = 0$ for $|z_1| = 1/2$, also $\tilde{u}(z_1,0,\ldots,0) = 0$ for $|z_1| \le 1/2$. Hence $\tilde{u}(z_1,0,\ldots,0) \le w(z_1)$, and since $w(z_1)$ is a subsolution corresponding to u , we obtain

$$\tilde{u}(z_1,0,\ldots,0) = w(z_1) , \quad |z_1| \le 1 .$$

In particular, $\tilde{u}$ is not twice differentiable. This example is taken from [5].

We remark in closing that the Bremermann functions can be thought of as generalized solutions to a special case of the complex Monge-Ampère equations. These equations are studied by E.Bedford and B.A.Taylor in [1], where a minimum principle is obtained that overlaps with the maximum principle of Theorem 5.19. The Dirichlet problem associated with the complex Monge-Ampère equations is to find a plurisubharmonic function u on D such that u takes on prescribed boundary values on ∂D , while $\det(H_u(z))$ is prescribed on D . The Bremermann functions are those solutions for which $\det(H_u(z)) = 0$ on D .

References

1. Bedford, E. and Taylor, B.A. The Dirichlet problem for a complex Monge-Ampère equation, *Inventiones Math.* 37 (1976), 1-44.
2. Bremermann, H. On a generalized Dirichlet problem for plurisubharmonic functions and pseudo-convex domains. Characterization of Shilov boundaries, *Trans. A.M.S.* 91 (1959), 246-276.
3. Diederich, K. and Fornaess, J.E., Pseudoconvex domains: bounded strictly plurisubharmonic exhaustion functions, *Inventiones Math.* 39 (1977), 129-141.

4. Freeman, M. Local complex foliation of real submanifolds, *Math. Annalen* 209 (1974), 1-30.
5. Gamelin, T.W. and Sibony, N. *Subharmonicity for uniform algebras*, to appear.
6. Hakim, M. and Sibony, N. Frontière de Šilov et spectre de $A(\bar{D})$ pour des domaines faiblement pseudoconvexes, *C. R. Acad. Sc. Paris* 281 (1975), 959-962.
7. Hormander, L. *Complex Analysis in Several Variables*, North Holland/American Elsevier, New York, 1973.
8. Rossi, H. Holomorphically convex sets in several complex variables, *Ann. Math.* 74 (1961), 470-493.
9. Vladimirov, V.S. *Methods of the Theory of Functions of Several Complex Variables*, M.I.T. Press, Cambridge, Mass., 1966.
10. Walsh, J.B. Continuity of envelopes of plurisubharmonic functions, *J. Math. Mech.* 18 (1968), 143-148.

Added in proof: In the development of the notion of subharmonicity with respect to an algebra of functions, the following reference also plays a key role.

11. Rickart, C.E. Plurisubharmonic functions and convexity properties for general function algebras, *Trans. A.M.S.* 169 (1972), 1-24.

7 · The conjugation operation for representing measures

There are a number of classical inequalities involving a trigonometric polynomial u and its conjugate polynomial $*u$. The prototypal result, due to M.Riesz, asserts that there exist constants c_p such that

$$\int |*u(e^{i\theta})|^p d\theta \leq c_p \int |u(e^{i\theta})|^p d\theta , \quad 1 < p < \infty . \tag{7.1}$$

Another important inequality is due to A.Zygmund:

$$\int |*u| d\theta \leq A \int |u| \log^+ |u| d\theta + B . \tag{7.2}$$

Both of these inequalities can be expressed in the form

$$\int H(u, *u) d\theta \geq 0 , \tag{7.3}$$

where H is a real-valued function on $\mathbb{C}$. Our aim is to study inequalities of the form (7.3), in a function algebra setting. We begin by introducing the conjugation operator.

The Conjugation Operator

Let A be a uniform algebra on X, and let $\phi \in M_A$. Let σ be a representing measure on X for ϕ. For $u \in \mathrm{Re}(A)$, let $*u$ denote an element of $\mathrm{Re}(A)$ such that $u + i*u \in A$ and $(*u)(\phi) = 0$. If v is another such element of $\mathrm{Re}(A)$, then $*u - v \in A$, and $(*u-v)(\phi) = 0$. Consequently

$$0 = (*u-v)^2(\phi) = \int (*u-v)^2 d\sigma ,$$

so that $v = {*u}$ a.e. $(d\sigma)$. It follows that ${*u}$ is determined uniquely by u and ϕ , at least on a set of full measure for each representing measure for ϕ . The function ${*u}$ is the *conjugate function* of u , and the operator $u \to {*u}$ is the *conjugation operator*.

Since $u + i{*u} - u(\phi)$ vanishes at ϕ , we have

$$0 = \mathrm{Re} \int (u + i{*u} - u(\phi))^2 d\sigma$$

$$= \int u^2 d\sigma - 2u(\phi) \int u d\sigma + u(\phi)^2 - \int ({*u})^2 d\sigma .$$

Substituting $\int u d\sigma = u(\phi)$, we are led to the identity

$$\int ({*u})^2 d\sigma + u(\phi)^2 = \int u^2 d\sigma . \tag{7.4}$$

In particular,

$$\int ({*u})^2 d\sigma \le \int |u|^2 d\sigma ,$$

so that the conjugation operator is bounded with respect to the norm of $L^2(\sigma)$.

More generally, the M.Riesz inequality is valid in this general setting, providing p is an even integer.

7.1 Theorem. Let σ be a representing measure for ϕ , and let p be an even integer. Then there is a constant c_p such that

$$\int |{*u}|^p d\sigma \le c_p \int |u|^p d\sigma , \qquad u \in \mathrm{Re}(A) . \tag{7.5}$$

Proof. Write $p = 2m$, where m is a positive integer. The polynomial $s^{2m} - 2(-1)^m \mathrm{Re}(1 + is)^{2m}$ has leading term $-s^{2m}$. Hence the polynomial is bounded above on the real axis, say

$$s^{2m} - 2(-1)^m \mathrm{Re}(1+is)^{2m} \le c , \qquad -\infty < s < \infty .$$

Substituting $s = y/x$, we are led to the estimate

$$y^{2m} \le cx^{2m} + 2(-1)^m \mathrm{Re}(x+iy)^{2m} , \qquad x,y \in \mathbb{R} .$$

Let $u \in \mathrm{Re}(A)$. Substituting u and $*u$ for x and y , and integrating with respect to σ , we obtain

$$\int (*u)^{2m} d\sigma \le c \int u^{2m} d\sigma + 2(-1)^m \mathrm{Re} \int (u+i*u)^{2m} d\sigma . \qquad (7.6)$$

If m is odd, we obtain the estimate (7.5) immediately. If m is even, we have

$$\int (u+i*u)^{2m} d\sigma = u(\phi)^{2m} = \left(\int u d\sigma\right)^{2m} \le \int u^{2m} d\sigma . \qquad (7.7)$$

Substituting (7.7) into (7.6), we obtain the estimate (7.5).

□

In the case that A is the disc algebra, one can invoke the M.Riesz Interpolation Theorem, to conclude that the estimate (7.1) is valid for $2 \le p < \infty$. By duality, the estimate (7.1) is also valid for $1 < p \le 2$, and the M.Riesz Theorem is proved. Moreover, it is easy to obtain bounds for the constants c_p appearing in (7.1).

Cole's Theorem

It turns out that the M.Riesz Theorem simply fails in the general setting, unless p is an even integer. That will be a consequence of the following theorem, due to B.Cole.

7.2 Theorem. Let D be a domain in the complex plane, let

$z_0 \in D$, and let H be a continuous real-valued function on D . Then the following are equivalent:

(i) There exists an analytic function F on D such that

$$\mathrm{Re}(F) \leq H$$

and

$$F(z_0) \geq 0 .$$

(ii) If A is a uniform algebra, $\phi \in M_A$, and σ is a representing measure for ϕ , then

$$\int (H \circ f) d\sigma \geq 0$$

for all $f \in A$ satisfying $f(M_A) \subset D$ and $f(\phi) = z_0$.

Proof. Suppose (i) is valid. Let A, ϕ, σ and f be as in (ii). Since F is analytic in a neighbourhood of $f(M_A)$, $F \circ f$ belongs to A and

$$\int (F \circ f) d\sigma = (F \circ f)(\phi) = F(z_0) \geq 0 .$$

Since

$$\int (F \circ f) d\sigma \leq \int (H \circ f) d\sigma ,$$

we obtain (ii). Observe that the proof of the M.Riesz Theorem given earlier is essentially a special case of this proof, in which F is the polynomial $2(-1)^m z^{2m}$.

Suppose now that (ii) is valid. Let $\{K_n\}_{n=1}^{\infty}$ be an increasing sequence of compact subsets of D such that $z_0 \in K_0$, and $\cup K_n = D$. Applying the hypothesis to the algebra $R(K_n)$ and the coordinate function z , we find that

$$\inf\{\int H d\sigma : \sigma \text{ represents } z_0 \text{ on } R(K_n)\} \geq 0 .$$

From Theorem 2.1 we conclude that the supremum of $Re(f(z_0))$ over all $f \in R(K_n)$ satisfying $Re(f) < H$ on K_n, is non-negative. Hence there is a function f_n analytic in a neighbourhood of K_n such that $Re(f_n) < H$ on K_n, and $f_n(z_0) > -1/n$. Any limit F of the normal family $\{f_n\}$ as $n \to \infty$ has the properties listed in (i).

□

7.3 Corollary. Let H be a continuous real-valued function on the plane. Then the following are equivalent:

(i) There is an entire function F on C such that

$$Re(F) \leq H$$

and

$$F(0) \geq 0 .$$

(ii) If A is a uniform algebra, $\phi \in M_A$, and σ is a representing measure for ϕ, then

$$\int H(u,{}^*u)\,d\sigma \geq 0$$

for all $u \in Re(A)$ satisfying $u(\phi) = 0$.

The M.Riesz Estimate

As a first application, we show that the version of the M.Riesz Theorem given in Theorem 7.1 fails in the case that p is not an even integer.

7.4 Theorem. Suppose that $1 \leq p \leq \infty$, and p is not an even integer. Then there is a uniform algebra A, and a representing measure σ for $\phi \in M_A$, such that the conjugation operator $u \to *u$ is not bounded in the norm of $L^p(\sigma)$.

Proof. Suppose that there is a constant $c > 0$ such that

$$\int |*u|^p d\sigma \le c \int u^p d\sigma \tag{7.8}$$

for all $u \in \mathrm{Re}(A)$ satisfying $u(\phi) = 0$. According to Corollary 7.3, there exists an entire function F such that

$$\mathrm{Re}(F) \le c|x|^p - |y|^p , \tag{7.9}$$

and

$$F(0) \ge 0 . \tag{7.10}$$

For fixed $t > 0$, the function

$$F_t(z) = F(tz)/t^p$$

also satisfies (7.9) and (7.10). Consequently the F_t's form a normal family. If

$$F(z) = a_0 + a_1 z + a_2 z^2 + \ldots ,$$

then the coefficient of z^m in the power series expansion of F_t is $a_m t^{m-p}$. By normality, these remain bounded as t ranges over the interval $(0,\infty)$. Hence $a_m = 0$ unless $m = p$. Since F cannot be identically zero, by (7.9), we conclude that p is an integer, and

$$F(z) = az^p .$$

The estimate (7.9) on the imaginary axis becomes

$$\mathrm{Re}(ai^p)y^p \le -|y|^p , \quad -\infty < y < \infty .$$

For this to hold, the integer p must be even.

Now suppose that p is not an even integer. We have shown that for each constant $c > 0$, there exists a uniform algebra A_c on a compact space X_c, a representing measure σ_c for a homomorphism $\phi_c \in M_{A_c}$, and a function $u_c \in \mathrm{Re}(A)$ such that

$$\int |{*u_c}|^p d\sigma_c > c \int |u_c|^p d\sigma_c . \tag{7.11}$$

Let $X = \Pi\{X_c : c > 0\}$, let π_c be the projection of X onto X_c, and let A be the uniform algebra on X generated by the functions $g\circ\pi_c$, where $g \in A_c$ and $c > 0$. Then the product σ of the measures σ_c is multiplicative on A, so that σ is a representing measure for some $\phi \in M_A$. Furthermore $U_c = u_c\circ\pi_c$ belongs to $\mathrm{Re}(A)$, and ${*U_c} = {*u_c}\circ\pi_c$. The estimate (7.11) shows that

$$\int |{*U_c}|^p d\sigma > c \int |U_c|^p d\sigma .$$

Hence the conjugation operator is not bounded on $\mathrm{Re}(A)$, in the norm of $L^p(\sigma)$. □

As a generalization of the preceding theorem, we prove the following.

7.5 Theorem. Suppose $0 < p < \infty$ and $0 < r < \infty$. The conjugation operator $u \to {*u}$ is continuous, from the L^p-norm to the L^r-norm, for all uniform algebras A, all $\phi \in M_A$, and all representing measures σ for ϕ, if and only if there is an even integer $2m$ such that $r \le 2m \le p$.

Proof. If $r \leq 2m \leq p$, the continuity of the conjugation operator from L^p to L^r follows from Theorem 7.1 and Hölder's inequality.

Assume conversely that the conjugation operator is continuous. We consider only the case $r \leq p$. We may assume that there exists $c > 0$ such that for all A, ϕ and σ,

$$\left(\int |{*u}|^r d\sigma\right)^{1/r} \leq \left(c \int |u|^p d\sigma\right)^{1/p} \quad , \quad u \in Re(A) .$$

Setting $b = p/r \geq 1$, we obtain the equivalent estimate

$$\left(\int |{*u}|^r d\sigma\right)^b \leq c \int |u|^p d\sigma \quad , \quad u \in Re(A) .$$

The function t^b is convex, so that the tangent line to the graph of t^b lies beneath the graph:

$$bt + 1 - b \leq t^b \quad , \quad -\infty < t < \infty .$$

This leads to the estimate

$$b \int |{*u}|^r d\sigma + 1 - b \leq c \int |u|^p d\sigma \quad , \quad u \in Re(A) .$$

By Corollary 7.3, there exists an entire function F such that $F(0) \geq 0$ and

$$Re(F(z)) \leq c|x|^p + b - 1 - b|y|^r \quad , \quad z = x + iy \in \mathbb{C} . \quad (7.12)$$

For $0 < t < \infty$ the entire function $F_t(z) = F(tz)/t^p$ satisfies

$$Re(F_t(z)) \leq c|x|^p + b/t^p$$

and

$F_t(0) \geq 0$.

It follows that the F_t's form a normal family as $t \to +\infty$. If $F(z) = \Sigma a_k z^k$ is the power series expansion of F , then the coefficient of z^k in the power series expansion of F_t is $a_k t^{k-p}$. By normality, these remain bounded as $t \to +\infty$. It follows that $a_k = 0$ for $k > p$, and F is a polynomial whose degree does not exceed p .

Replacing F by $[F(z) + \overline{F(\bar{z})}]/2$, we can assume that the coefficients of F are real. Furthermore, replacing F by $[F(z) + F(-z)]/2$, we can assume that F is an even polynomial. In particular, the degree of F is an even integer $2m$, and $2m \leq p$. Along the imaginary axis $F(iy)$ is asymptotic to $(-1)^m a_{2m} y^{2m}$ as $|y| \to +\infty$. Comparing this with (7.12), we obtain $r \leq 2m$. □

7.6 Corollary. If $0 < p < 1$, then the Kolmogoroff estimate

$$\left(\int |*u|^p d\sigma\right)^{1/p} \leq c \int |u| d\sigma \quad , \quad u \in \mathrm{Re}(A) ,$$

fails for some uniform algebra A and some representing measure σ . Furthermore, the Zygmund estimate

$$\int |*u| d\sigma \leq \alpha + \beta \int |u| \log^+ |u| d\sigma \quad , \quad u \in \mathrm{Re}(A)$$

fails for some A and σ .

Proof. The first statement follows immediately from Theorem 7.5. Since

$$\int |u| \log^+ |u| d\sigma \leq \int |u|^{3/2} d\sigma \quad ,$$

the Zygmund estimate is valid only when the conjugation operator is continuous from the $L^{3/2}$-norm to the L^1-norm.

This also fails in general, by the preceding theorem.

□

If we restrict our attention to positive functions in $Re(A)$, then the M.Riesz Theorem remains valid, at least in the case that p is not an odd integer.

7.7 Theorem. Let A be a uniform algebra, and let σ be a representing measure for some $\phi \in M_A$. Suppose that $1 < p < \infty$, and that p is not an odd integer. Then there is a constant c_p such that

$$\int |{*u}|^p d\sigma \le c_p \int |u|^p d\sigma$$

for all $u \in Re(A)$ satisfying $u > 0$. The best possible constants c_p tend to $+\infty$ as p tends to an odd integer.

Proof. Choose $\gamma > 0$ and α real such that

$$\alpha \cos(p\theta) \le \gamma|\cos\theta|^p - |\sin\theta|^p \quad , \qquad 0 \le \theta \le \pi/2 . \qquad (7.13)$$

To see that this choice is possible, observe that $\cos(p\,\pi/2) \ne 0$, since p is not an odd integer. Hence we can choose α, possibly negative, so that (7.13) is valid near $\theta = \pi/2$, and then we can choose $\gamma > 0$ so large that the estimate is valid on the remainder of the interval. In terms of the principal branch of the function αz^p, the estimate (7.13) becomes

$$Re(\alpha z^p) \le \gamma|x|^p - |y|^p \quad , \qquad (7.14)$$

where $z = x + iy$ belongs to the right half-plane $\{x > 0\}$. Now there are two cases to consider.

If $\alpha > 0$, then αz^p is positive on the positive real axis. Applying Theorem 7.2, with D the right half-plane and $z_0 = \int u d\sigma$, we obtain

$$0 \le \gamma \int |u|^p d\sigma - \int |{*u}|^p d\sigma$$

for all $u \in \mathrm{Re}(A)$ such that $u > 0$. A slightly better estimate

$$\alpha \left(\int u d\sigma \right)^p \le \gamma \int |u|^p d\sigma - \int |{*u}|^p d\sigma$$

is obtained by substituting $x = u$ and $y = {*u}$ in (7.14) and integrating. In any event, we obtain the required estimate in the case at hand.

Suppose on the other hand that $\alpha < 0$. We will use the estimate

$$px - p + 1 \le x^p , \tag{7.15}$$

which reflects the fact that the tangent line to the graph of x^p at $x = 1$ lies beneath the graph. From (7.14) and (7.15) we see that the function

$$F(z) = \alpha z^p - \alpha(pz - p + 1)$$

satisfies

$$\mathrm{Re}\, F(z) \le (\gamma-\alpha)|x|^p - |y|^p$$

in the right half-plane. Note also that $F(1) = 0$. Applying Theorem 7.2, with D the right half-plane and $z_0 = 1$, we obtain

$$0 \le (\gamma-\alpha) \int |u|^p d\sigma - \int |{*u}|^p d\sigma$$

for all $u \in \mathrm{Re}(A)$ satisfying $u > 0$ and $\int u\,d\sigma = 1$. Since the inequality is homogeneous, we can drop the condition $\int u\,d\sigma = 1$, and we obtain the desired estimate.

It remains to prove the final assertion of the theorem. We do this by means of an example, in which σ is a representing measure for 0 with respect to the disc algebra $A(\Delta)$.

Fix an odd integer $2n+1$, where $n \geq 1$. Define a function h on $\partial\Delta$ by

$$h(e^{i\theta}) = |e^{i\theta} - 1|^{2n} = \sum_{k=-n}^{n} d_k e^{ik\theta} .$$

We extend h to $\mathbb{C}$ to be a harmonic polynomial, by setting

$$h(re^{i\theta}) = \sum_{-n}^{n} d_k r^{|k|} e^{ik\theta} .$$

Fix $0 < a < 1$, and let η be the normalized arc-length measure on the circle $\partial\Delta_a = \{|z| = a\}$, so that η represents 0. Define g on $\partial\Delta_a$ by

$$g(ae^{i\theta}) = h(\frac{1}{a} e^{i\theta}) .$$

We claim that $g\eta - h\,d\theta/2\pi$ is orthogonal to all harmonic polynomials. Indeed, if $f = \Sigma c_k r^{|k|} e^{ik\theta}$ is a harmonic polynomial, then

$$\int f(e^{i\theta}) h(e^{i\theta}) \frac{d\theta}{2\pi} = \frac{1}{2\pi} \int (\sum c_j e^{ij\theta})(\sum d_k e^{ik\theta}) d\theta$$

$$= \sum c_j d_{-j} ,$$

and this coincides with

$$\int f(ae^{i\theta}) g(ae^{i\theta}) d\eta = \frac{1}{2\pi} \int (\sum c_j a^{|j|} e^{ij\theta})(\sum d_k a^{-|k|} e^{ik\theta}) d\theta .$$

Now choose $\varepsilon > 0$ so that $1 - \varepsilon g$ is strictly positive on $\partial\Delta_a$, and define

$$\sigma = (1 - \varepsilon g)\eta + \varepsilon h \frac{d\theta}{2\pi} = \eta + \varepsilon(h \frac{d\theta}{2\pi} - g\eta) .$$

Since $h \geq 0$ on $\partial\Delta$, σ is a positive measure. Since $g\eta - hd\theta/2\pi$ is orthogonal to analytic polynomials, σ is a representing measure for 0.

Define $F_\delta \in A(\Delta)$, for $\delta > 0$, by

$$F_\delta = (1 + z)/(1 + \delta - z) = u_\delta + iv_\delta .$$

Then $u_\delta > 0$, so that $c_p \geq \int |v_\delta|^p d\sigma / \int |u_\delta|^p d\sigma$. If $p < 2n + 1$, then F_δ converges in $L^p(\sigma)$ as δ decreases to 0, to

$$F = (1+z)/(1-z) = u + iv .$$

Here we have used the dominated convergence theorem, and the fact that h has a zero of order $2n$ at $z = 1$. Hence $c_p \geq \int |v|^p d\sigma / \int |u|^p d\sigma$, and one computes, noting that $u = 0$ on $\partial\Delta$, that this latter quantity is on the order of

$$\frac{\int |1 - e^{i\theta}|^{2n-p} d\theta + \int |v(ae^{i\theta})|^p [1 - \varepsilon g(ae^{i\theta})] d\theta}{\int |u(ae^{i\theta})|^p [1 - \varepsilon g(ae^{i\theta})] d\theta} .$$

This tends to $+\infty$ as p increases to $2n + 1$.

□

Estimates of Zygmund

While the Zygmund estimate fails in general, it is valid for functions with positive real part.

7.8 Theorem. For each $\alpha > 2/\pi$, there exists $\beta > 0$ such that

$$\int |{*u}|\,d\sigma \le \beta + \gamma \int u \log^+ u \, d\sigma$$

for all uniform algebras A, all representing measures σ, and all $u \in \mathrm{Re}(A)$ such that $u > 0$.

Proof. Let $\delta < \beta$ satisfy $\delta > 2/\pi$. The estimate

$$|\sin\theta| \le \delta \cos\theta \log(\cos\theta) + \delta\theta \sin\theta$$

is valid near $\theta = \pm\pi/2$. Consequently we can choose $\alpha > 0$ so large that

$$|\sin\theta| \le \delta\cos\theta\log(\cos\theta) + \delta\theta\sin\theta + \alpha\cos\theta\,, \quad |\theta| \le \pi/2\,.$$

This leads to the estimate

$$\begin{aligned}\mathrm{Re}(\delta z \log z - \alpha z) &= \delta r \cos\theta \log r - \delta r\theta \sin\theta - \alpha r \cos\theta \\ &\le \delta r \cos\theta \log r + \delta r \cos\theta \log(\cos\theta) - r\,|\sin\theta| \\ &= \delta x \log x - |y|\,,\end{aligned}$$

for $z = x + iy$ and $x > 0$. Substituting $x = u$ and $y = {*u}$, and integrating, we obtain

$$\delta u(\phi) \log u(\phi) - \alpha u(\phi) \le \delta \int u \log u \, d\sigma - \int |{*u}|\,d\sigma\,.$$

Since $t \log t \ge -1$ for $t > 0$, we obtain

$$\int |{*u}|\,d\sigma \le \delta + \delta \int u \log u\,d\sigma + \alpha u(\phi)\,. \tag{7.16}$$

For any $\varepsilon > 0$, there exists $c > 0$ such that

$$u(\phi) = \int u d\sigma \le \varepsilon + c \int u \log^+ u \, d\sigma .$$

Combining this estimate with (7.16), and choosing ε so that $\delta + \alpha\varepsilon = \beta$, we obtain the Zygmund inequality.

□

There is a converse estimate to Theorem 7.8, namely, that

$$\frac{2}{\pi} \int u \log u \, d\sigma \le \int |{*u}| d\sigma + \frac{2}{\pi} u(\phi) \log u(\phi) \tag{7.17}$$

whenever σ represents ϕ and $u \in Re(A)$ is positive. To prove this, observe that

$$\begin{aligned} Re(-z \log z) &= - r \cos\theta \log r + r\theta \sin\theta \\ &= - x \log x + \theta y + r \cos\theta \log(\cos\theta) \\ &\le - x \log x + \frac{\pi}{2} |y| . \end{aligned}$$

Substituting $z = u + i{*u}$ and integrating, we obtain

$$-u(\phi) \log u(\phi) \le - \int u \log u \, d\sigma + \frac{\pi}{2} \int |{*u}| d\sigma ,$$

which is the same as (7.17). Since $t \log t \ge -1/e$, (7.17) yields the estimate

$$\int u \log^+ u \, d\sigma \le \frac{\pi}{2} \int |{*u}| d\sigma + \frac{1}{e}$$

for all $u \in Re(A)$ satisfying $u > 0$ and $u(\phi) = 1$.

There is another estimate, that is also due to Zygmund.

7.9 Theorem. Let A be a uniform algebra, and let σ be a representing measure for $\phi \in M_A$. If $u \in Re(A)$ satisfies $|u| \le 1$, and if $0 \le \alpha < \pi/2$, then

$$\int e^{\alpha|*u|}d\sigma \le 2/\cos\alpha \ .$$

Proof. In this case, we take D to be the vertical strip $\{-1 < \text{Re}(z) < +1\}$, and we consider

$$F(z) = 2 - e^{i\alpha z} - e^{-i\alpha z} = 2(1-\cos(\alpha z)) \ .$$

Note that $F(x) > 0$ for $-1 < x < 1$. The estimates

$$e^{\pm\alpha y}\cos\alpha \le e^{\pm\alpha y}\cos(\alpha x) = \text{Re}(e^{\mp i\alpha z})$$

combine to yield

$$e^{\alpha|y|}\cos\alpha \le (e^{\alpha y} + e^{-\alpha y})\cos\alpha \le \text{Re}(e^{i\alpha z} + e^{-i\alpha z}) \ .$$

Hence

$$\text{Re}(F(z)) \le 2 - e^{\alpha|y|}\cos\alpha \quad , \quad -1 < \text{Re}(z) < 1 \ .$$

This leads to the estimate

$$0 \le 2 - \cos\alpha \int e^{\alpha|*u|}d\sigma$$

whenever $u \in \text{Re}(A)$ satisfies $-1 < u < 1$.

□

The Kolmogoroff Estimate

The Kolmogoroff estimate is easiest of all.

7.10 Theorem. If $0 < p < 1$, then

$$\int |*u|^p d\sigma \le \frac{1}{\cos(\frac{p\pi}{2})}\left(\int u d\sigma\right)^p \tag{7.18}$$

for all uniform algebras A , all representing measures σ , and all $u \in Re(A)$ satisfying $u > 0$.

Proof. The estimates

$$\cos(\frac{p\pi}{2})\ |y|^p \le \cos(\frac{p\pi}{2})r^p \le \cos(p\theta)r^p = Re(z^p)$$

yield

$$Re(1 - z^p) \le 1 - \cos(\frac{p\pi}{2})\ |y|^p \ .$$

Applying Theorem 7.2, with $F(z) = 1 - z^p$, we obtain

$$0 \le 1 - \cos(\frac{p\pi}{2}) \int |{*u}|^p d\sigma$$

whenever $u \in Re(A)$ satisfies $u > 0$ and $\int u d\sigma = 1$. This yields (7.18) in the case $\int u d\sigma = 1$. Since the inequality (7.18) is homogeneous, the restriction $\int u d\sigma = 1$ is unnecessary. □

Weak-type Estimates

Next we treat weak-type estimates, showing first that they fail in general, then that they obtain providing $u \in Re(A)$ is positive.

7.11 Theorem. Let $0 < \alpha < 1$, $\beta > 0$ and $\lambda > 0$. Then the estimate

$$\sigma(\{|{*u}| > \lambda\}) \le \alpha + \beta \int |u| d\sigma \qquad (7.19)$$

fails for some uniform algebra A and some representing measure σ .

Proof. Suppose (7.19) is valid for all A and σ . By Corollary 7.3, there is an entire function F such that

$F(0) \geq 0$, and

$$\mathrm{Re}(F(z)) \leq \alpha + \beta|x| - \chi , \tag{7.20}$$

where χ is the characteristic function of the set $\{z : |\mathrm{Im}(z)| > \lambda\}$. (Note that Corollary 7.3 is valid also for upper semi-continuous functions. This can be seen by approximating such a function from above by continuous functions.)

For fixed $t > 0$, the entire function $F_t(z) = F(tz)/t$ satisfies

$$\mathrm{Re}(F_t(z)) \leq \frac{\alpha}{t} + \beta .$$

Again the family $\{F_t : t \geq 1\}$ is normal. The boundedness of the Taylor coefficients of F_t as t tends to ∞ shows that F is linear, say $F(z) = a + bz$, where $a \geq 0$. Substituting $z = iy$ into (7.20), we obtain

$$a + y\,\mathrm{Re}(ib) \leq \alpha - 1 \quad , \qquad |y| > \lambda .$$

Since $\alpha - 1 < 0$, this is impossible.

□

7.12 Theorem. If A is a uniform algebra, and σ is a representing measure for A , then

$$\sigma(\{|{*u}| \geq \lambda\}) \leq \frac{4}{\pi\gamma} \int u\,d\sigma \quad , \qquad \lambda > 0 , \tag{7.21}$$

for all $u \in \mathrm{Re}(A)$ satisfying $u > 0$.

Proof. Consider the harmonic function

$$U(z) = 1 + \frac{1}{\pi}\arg(z - i\lambda) - \frac{1}{\pi}\arg(z + i\lambda)$$

on the right half-plane. Its range lies in the interval $[0,1]$, it is zero on the interval $(-i\lambda, i\lambda)$, and it is 1 on the intervals $(-i\infty, -i\lambda)$ and $(i\lambda, i\infty)$. Furthermore, $U \geq \chi/2$ on the right half-plane, and

$$U(1) = 1 - \frac{2}{\pi} \text{arc tan } \lambda \leq 2/(\pi\lambda) .$$

Hence the function

$$F(z) = \frac{4z}{\pi\lambda} - 1 + \frac{i}{\pi} \log(\frac{z - i\lambda}{z + i\lambda})$$

satisfies

$$\text{Re } F \leq \frac{4x}{\pi\lambda} - \chi$$

and $F(1) \geq 0$. We conclude that

$$0 \leq \frac{4}{\pi\lambda} \int u d\sigma - \sigma(\{|{*u}| \geq \lambda\}) \tag{7.22}$$

whenever $u \in \text{Re}(A)$ satisfies $u > 0$ and $\int u d\sigma = 1$. The general case (7.21) is obtained from (7.22) by replacing $u + i{*u}$ by $(u + i{*u})/t$ and λ by λ/t, where $t = \int u d\sigma$.

□

Notes on Sources

We close with some comments on the origin of the material in this section.

The approach that we have followed is that laid out by B.Cole in a colloquium talk at Tulane in December, 1970. In particular, Cole obtained Theorem 7.2 and used it to show that many of the classical estimates fail for general representing measures. The story will be continued in the next lecture, which deals with Jensen measures.

The M.Riesz inequality was announced in 1924 [9]. According

to Riesz, he prepared the details for publication in that year, but then he delayed submission of the manuscript for two years, so that the proofs appeared only in 1927 [11]. The 1927 paper of Riesz is a classic. We mention several of the highlights.

First Riesz obtains the estimate (7.1) for even integers, with a proof along roughly the same lines as the proof of Theorem 7.1. (Our proof of Theorem 7.1 is Cole's simplification of Riesz's proof.) Riesz goes on to give a proof of (7.1) in case p is not an odd integer, that is based on contour integration. He then handles the exceptional cases by duality.

Riesz returns to the proof covering even integers in order to estimate the constants c_p, and he finds that $c_{2m} = O(m)$ as $m \to +\infty$. He remarks that it would be interesting to determine how the best possible constant depends on p, and in a footnote "added in proof", he cites his paper [10] of 1926, in which he obtains his celebrated convexity theorem. Riesz did not use the convexity theorem to obtain his estimates on conjugate harmonic functions, but rather he was apparently led to the convexity theorem in seeking to understand his estimates.

The idea of basing the proof of the M.Riesz Theorem on the estimate (7.13) is due to A.P.Calderón[3].

It was Bochner[2] who observed that Riesz's proof for the case of even integers extends to a uniform algebra setting. Except for the context, Bochner's proof is identical to that of Riesz. Bochner was apparently unaware of this, and he omits reference to the Riesz paper.

An example in which the M.Riesz estimate of Theorem 7.7 fails, in the case $p = 3$, was given by K.Yabuta[12]. The example we have given, covering all odd integers, is due to H.König[6]. The example sheds light on the failure at precisely the odd integers of the complex-variable technique of

M.Riesz, in proving his classical estimate.

The estimate of Theorem 7.8 is due to Zygmund[13]. A proof based on contour integration was obtained by J.E. Littlewood[8], and the proof given here is due to Calderón [3]. It was apparently M.Riesz who observed (cf. [15, vol. I, p.381]) that the converse of the Zygmund estimate is valid, so that in particular if u and $*u$ belong to $L^1(d\theta)$, and if $u > 0$, then $u \log^+ u \in L^1(d\theta)$.

The estimate of Theorem 7.9 is also due to Zygmund[14].

The weak-type estimate of Theorem 7.12 and the estimate of Theorem 7.10 are due to A.N.Kolmogoroff[5]. In [5], Kolmogoroff first obtained the weak-type estimate, and he deduced from this the boundedness of the conjugation operator from $L^1(d\theta)$ to $L^p(d\theta)$, $0 < p < 1$. Littlewood[7] gave a proof of the Kolmogoroff estimate using complex variable techniques, and this proof was simplified by G.H.Hardy[4] to the now standard proof.

Strictly speaking, Kolmogoroff's weak-type estimate precedes the other estimates we have considered. His results were submitted for publication early in 1923. It should be noted though that A.Besicovitch[1] had already obtained a weak-type estimate for the Hilbert transform.

References

1. Besicovitch, A. Sur la nature des fonctions à carré sommable mesurables, *Fund. Math.* 4 (1923), 172-195.
2. Bochner, S. Generalized conjugate and analytic functions without expansions, *Proc. Nat. Acad. Sci.* 44 (1959), 855-857.
3. Calderón, A.P. On theorems of M.Riesz and Zygmund, *Proc. A.M.S.* 1 (1950), 533-535.
4. Hardy, G.H. Remarks on three recent notes in the Journal, *J. London Math. Soc.* 3 (1928), 166-169.
5. Kolmogoroff, A.N. Sur les fonctions harmoniques

conjugées et les séries de Fourier, *Fund. Math.* 7 (1925), 23-28.

6. König, H. On the Marcel Riesz estimation for conjugate functions in the abstract Hardy theory, *Commentations Math.* (1978).
7. Littlewood, J.E. On a theorem of Kolmogoroff, *J. London Math. Soc.* 1 (1926), 229-231.
8. Littlewood, J.E. On a theorem of Zygmund, *J. London Math. Soc.* 4 (1929), 305-307.
9. Riesz, M. Les fonctions conjugées et les séries de Fourier, *C. R. Acad. Sci. Paris* 178 (1924), 1464-1467.
10. Riesz, M. Sur les maxima des formes bilinéaires et sur les fonctionnelles linéaires, *Acta Math.* 49 (1926), 456-497.
11. Riesz, M. Sur les fonctions conjugées, *Math. Zeitschrift* 27 (1927), 218-244.
12. Yabuta, K. M.Riesz's theorem in the abstract Hardy space theory, *Arch. Math.* 29 (1977), 308-312.
13. Zygmund, A. Sur les fonctions conjugées, *Fund. Math.* 13 (1929), 284-303.
14. Zygmund, A. *Trigonometric Series*, 2nd ed., Cambridge University Press, 1968.

8 · The conjugation operation for Jensen measures

While the M.Riesz and Zygmund estimates fail in general, they turn out to be valid for Jensen measures, and the constants are the same as those that arise in the case of the disc algebra. This is a consequence of the implication "(iii) implies (i)" of Theorem 8.3, which is due to B.Cole. Before proving Cole's theorem, we present yet another proof of the M.Riesz Theorem, which will serve to illustrate the underlying idea.

The M.Riesz Estimate for Jensen Measures

Let us first consider the classical case.

Fix $1 < p < 2$. Define h on the right half-plane by

$$h(r \cos \theta, r \sin \theta) = r^p \cos(p\theta), \quad |\theta| \le \pi/2,$$

and extend h to $\mathbb{C}$ so that h is symmetric about the imaginary axis:

$$h(x,y) = h(-x,y), \quad x + iy \in \mathbb{C}.$$

Note that h is continuous, and that h is harmonic except on the imaginary axis. We claim that h is subharmonic.

Consider first the behaviour of h near the positive imaginary axis $\{\theta = \pi/2\}$. Since

$$\frac{\partial}{\partial\theta} r^p \cos(p\theta) = -pr^p \sin(p\theta)$$

has negative sign for $\theta = \pi/2$, the values of $r^p \cos(p\theta)$ are less than those of the reflected function $r^p \cos[p(\pi-\theta)]$

for $\frac{\pi}{2} < \theta < \frac{\pi}{2} + \varepsilon$, while $r^p \cos[p(\pi-\theta)]$ is dominated by $r^p \cos(p\theta)$ for $\frac{\pi}{2} - \varepsilon < \theta < \frac{\pi}{2}$. It follows that h is the maximum of the harmonic functions $r^p \cos(p\theta)$ and $r^p \cos [p(\pi-\theta)]$ near the positive imaginary axis. Consequently h is subharmonic there. Similarly, h is subharmonic near the negative imaginary axis. To check that h satisfies the mean value estimate at the origin, we simply compute

$$\int_{-\pi}^{\pi} h(r\cos\theta,\ r\sin\theta)d\theta = 2r^p \int_{-\pi/2}^{\pi/2} \cos(p\theta)d\theta$$
$$> 0 = h(0,0)\ .$$

Hence h is subharmonic.

As in the preceding chapter, we choose $\alpha > 0$ and $\gamma > 0$ such that

$$\alpha \cos(p\theta) \le \gamma|\cos\theta|^p - |\sin\theta|^p\ , \quad -\frac{\pi}{2} \le \theta \le \frac{\pi}{2}\ . \tag{8.1}$$

In terms of the subharmonic function h, the inequality becomes

$$\alpha h(x,y) \le \gamma|x|^p - |y|^p\ . \tag{8.2}$$

If u is a trigonometric polynomial, then $h(u,*u)$ is subharmonic, so that

$$|u(0)|^p = h(u(0),0) \le \int h(u,*u)\,\frac{d\theta}{2\pi}\ . \tag{8.3}$$

Substituting $x = u$ and $y = *u$ in (8.2), and integrating, we obtain

$$\alpha \int h(u,*u)\,\frac{d\theta}{2\pi} \le \gamma \int |u|^p\,\frac{d\theta}{2\pi} - \int |*u|^p\,\frac{d\theta}{2\pi}\ . \tag{8.4}$$

Now (8.3) and (8.4) lead to the estimate

$$\int |{*u}|^p \frac{d\theta}{2\pi} + \alpha|u(0)|^p \le \gamma \int |u|^p \frac{d\theta}{2\pi}, \tag{8.5}$$

which proves in particular the M.Riesz Theorem, for $1 < p < 2$.

Handling the range $2 < p < \infty$ by duality, we obtain what is currently the "best" proof of the classical M.Riesz Theorem.

For the range $1 < p < 2$, an appropriate modification of the above proof remains valid when $d\theta/2\pi$ is replaced by an arbitrary Jensen measure. In fact, the only step of the proof that does not extend immediately to arbitrary representing measures is the estimate $h(u(\phi),0) \le \int h(u,{*u})d\sigma$. In the case of Jensen measures, this estimate is a simple consequence of the results of Chapter 3. We state the result separately, for emphasis.

8.1 Lemma. If σ is a Jensen measure for $\phi \in M_A$, then

$$h(f(\phi)) \le \int h\circ f \, d\sigma$$

for all $f \in A$ and all functions h that are subharmonic in a neighbourhood of the range $f(M_A)$ of f on M_A.

Proof. Let $f*\sigma$ be the probability measure on $f(M_A)$ defined so that

$$\int w d(f*\sigma) = \int w\circ f \, d\sigma, \qquad w \in C(f(M_A)).$$

If g is a rational function with poles off $f(M_A)$, then $g\circ f$ belongs to A, so that

$$\log|g(f(\phi))| \le \int \log|g\circ f|d\sigma = \int \log|g|d(f*\sigma).$$

Hence $f*\sigma$ is a Jensen measure for $f(\phi)$ with respect to the algebra $R(f(M_A))$. The lemma now follows from

Theorem 3.4, applied to $\nu = f^*\sigma$ and $p = f(\phi)$.

□

Thus we see that subharmonic functions give rise to estimates for integrals, and the particular subharmonic function used above gives rise to the estimate

$$\int |{*u}|^p d\sigma + \alpha |u(\phi)|^p \le \gamma \int |u|^p d\sigma , \quad u \in \mathrm{Re}(A) ,$$

where σ is any Jensen measure for $\psi \in M_A$. Cole's theorem will tell us in addition that all integral estimates arise in the above manner from subharmonic functions.

Cole's Theorem

The proof of Cole's theorem depends on the following lemma from potential theory.

8.2 Lemma. Let K be a compact subset of $\mathbb{C}$ such that $\mathbb{C}\backslash K$ is connected, and let H be a continuous, real-valued function on K . Let h be the upper envelope of the family of functions that are subharmonic in a neighbourhood of K and are dominated by H on K . Then h is continuous, h is subharmonic on the interior of K , and $h = H$ on ∂K . Furthermore, h is harmonic on the open set $\{h < H\}$.

Sketch of proof. The fact that $h = H$ on ∂K follows from the fact that every point of ∂K is a "stable" boundary point, that is, a regular boundary point for the outer Dirichlet problem. Another way of looking at it, from the point of view of Chapter 5, is to observe that h is the solution $\tilde{H}$ of the $R(K)$-Dirichlet problem with boundary data H on all of K . In this case, each point of ∂K is a Jensen boundary point, so that $h = H$ on ∂K .

The crux of the proof involves showing that the set

$\{h < H\}$ is open. This depends on a standard estimate for harmonic measure, and the details are given in [4].

Once we know that $\{h < H\}$ is open, we deduce that h is harmonic on $\{h < H\}$, since it is the upper envelope of a Perron family of subharmonic functions. Since h is lower semi-continuous, it is continuous at each point of the set $\{h = H\}$, hence everywhere on K. Being the upper envelope of subharmonic functions, h is subharmonic on K^o.

□

8.3 Theorem *(Cole's Theorem)*. Let H be a continuous, real-valued function on $\mathbb{C}$. Then the following are equivalent.

(i) There exists a subharmonic function h on $\mathbb{C}$ such that $h \le H$, while $h \ge 0$ on the real axis.

(ii) If A is a uniform algebra, $\phi \in M_A$, and σ is a Jensen measure for ϕ, then

$$\int H(u,{*u})d\sigma \ge 0 , \qquad u \in \mathrm{Re}(A) .$$

(iii) For all trigonometric polynomials u,

$$\int_0^{2\pi} H(u(e^{i\theta}),{*u}(e^{i\theta}))d\theta \ge 0 .$$

Proof. Suppose first that (i) is valid. Let σ be a Jensen measure for $\phi \in M_A$, let $u \in \mathrm{Re}(A)$, and set $f = u+i{*u} \in A$. Using Lemma 8.1, we obtain $0 \le h(f(\phi)) \le \int h\circ f\, d\sigma \le \int H\circ f\, d\sigma$. The latter integral is the same as $\int H(u,{*u})d\sigma$, so that (ii) is proved.

Since (iii) is a special case of (ii), it remains to show that (iii) implies (i). For this, we suppose that (i) fails. We must show that (iii) also fails.

Let Δ_R be the closed disc $\{|z| \le R\}$, and let h_R denote the upper envelope of the functions that are subharmonic in a neighbourhood of Δ_R and are dominated by H on Δ_R.

By Lemma 8.2, h_R is continuous on Δ_R, and h_R is subharmonic on the interior of Δ_R.

As R tends to ∞, the functions h_R decrease to a subharmonic function h on $\mathbb{C}$ that is the greatest subharmonic minorant of H on $\mathbb{C}$. The hypothesis (i) is equivalent to the assertion that $h \geq 0$ on $\mathbb{R}$. Since (i) fails, there exist $R > 0$ and $x_0 \in \mathbb{R} \cap \Delta_R$ such that $h_R(x_0) < 0$.

If $H(x_0,0) < 0$, then (iii) fails for the constant function $u = x_0$. Hence we assume that

$$H(x_0,0) \geq 0 .$$

In particular, $h_R(x_0) < H(x_0,0)$.

Let U be the connected component of the set $\{z \in \Delta_R : h_R(z) < H(z)\}$ containing x_0. By Lemma 8.2, U is an open subset of the interior of Δ_R, and h_R is harmonic on U. Note that h_R is continuous on $\bar{U}$, and $h_R = H$ on ∂U.

Let π be the universal covering map of the open unit disc Δ onto U, normalized so that $\pi(0) = x_0$. Consider the functions $h_R \circ \pi$ and $H \circ \pi$ on Δ. Since the nontangential boundary values of π lie in ∂U, whenever they exist, we see that $h_R \circ \pi$ and $H \circ \pi$ have the same nontangential boundary values a.e. $(d\theta)$ on $\partial\Delta$. Since $h_R \circ \pi$ is harmonic on Δ, we have

$$0 > h_R(x_0) = (h_R \circ \pi)(0) = \int (h_R \circ \pi)(e^{i\theta}) \frac{d\theta}{2\pi}$$

$$= \int (H \circ \pi)(e^{i\theta}) \frac{d\theta}{2\pi} .$$

Approximating π in the weak-star topology of $H^\infty(d\theta)$ by analytic polynomials, we find an analytic polynomial g such that

$$\int (H \circ g)(e^{i\theta}) d\theta < 0 . \qquad (8.6)$$

Since $\pi(0)$ is real, we can assume that $g(0)$ is real. Then g takes the form $u + i{*u}$, and the inequality (8.6) becomes

$$\int H(u, {*u})d\theta < 0 .$$

Hence (iii) fails. □

Several variants of Theorem 8.3 are possible. For instance, one can consider an estimate of the form

$$\int H \circ f \, d\sigma \geq 0$$

for $f \in A$ such that $f(M_A)$ is included in some specified domain D and $f(\phi)$ in some specified subset $S \subset D$. The validity of this estimate for all Jensen measures is equivalent to the existence of a subharmonic function h on D such that $h \leq H$ on D, while $h \geq 0$ on S.

Theorem 8.3 remains valid if H is only upper semi-continuous. One merely approximates H from above by continuous functions.

With a slight modification, Theorem 8.3 is also valid in the case that H is a lower semi-continuous function from $\mathbb{C}$ to $(-\infty, +\infty]$. In this case, the function h of Theorem 8.3(i) is not necessarily subharmonic, but rather lower semi-continuous and quasi-subharmonic. In the terminology of Chapter 5, h is a log-envelope function.

Following Cole, we now deduce the various classical estimates for the conjugation operator in the case of a Jensen measure.

8.4 Theorem. For each $\beta > 2/\pi$, there exists $\gamma > 0$ such that for all uniform algebras A, $\phi \in M_A$ and Jensen measures σ for ϕ, we have

$$\int |{*u}|d\sigma \le \beta \int |u| \log^+|u|d\sigma + \gamma , \quad u \in \mathrm{Re}(A) . \tag{8.7}$$

Moreover, if $0 < p < 1$, there are constants c_p such that

$$\left(\int |{*u}|^p d\sigma \right)^{1/p} \le c_p \int |u|d\sigma , \quad u \in \mathrm{Re}(A) . \tag{8.8}$$

If $1 < p < \infty$, there are constants c_p such that

$$\left(\int |{*u}|^p d\sigma \right)^{1/p} \le c_p \left(\int |u|^p d\sigma \right)^{1/p} , \quad u \in \mathrm{Re}(A) . \tag{8.9}$$

There is a constant $c > 0$ such that if $\lambda > 0$, then

$$\sigma(\{|{*u}| \ge \lambda\}) \le \frac{c}{\lambda} \int |u|d\sigma , \quad u \in \mathrm{Re}(A) . \tag{8.10}$$

Furthermore, any constants which serve in these estimates for the disc algebra and the measure $d\theta/2\pi$ also serve for any uniform algebra and any Jensen measure.

Proof. Consider first the Zygmund estimate (8.7). In the case of the disc algebra, one writes $u = u_+ - u_-$, where $u_+ = \max(u,0)$ and $u_- = -\min(u,0)$. The estimate of Theorem 7.8 for close approximants of u_+ and u_- leads immediately to the estimate (8.7), for the disc algebra. The validity of the Zygmund estimate for a Jensen measure σ then follows from Theorem 8.3, since the estimate can be cast in the form $\int H(u,{*u})d\sigma \ge 0$.

The Kolmogoroff estimate (8.8) is established for the disc algebra as above. However, this estimate cannot be expressed in the form $\int H(u,{*u})d\sigma \ge 0$. We argue as follows.

Substituting $t = \int |{*u}|^p d\theta/2\pi$ in the inequality

$$\frac{1}{p} t + 1 - \frac{1}{p} \le t^{1/p} , \quad -\infty < t < +\infty , \tag{8.11}$$

we obtain

$$\frac{1}{p}\int |{*u}|^p d\sigma + 1 - \frac{1}{p} \le c_p \int |u| d\sigma \tag{8.12}$$

when $\sigma = d\theta/2\pi$ and u is a trigonometric polynomial. Now the estimate (8.12) can be written in the form $\int H(u,{*u})d\sigma \ge 0$. By Theorem 8.3, (8.12) is valid for all Jensen measures σ and all $u \in \mathrm{Re}(A)$. Since the inequality (8.11) becomes equality for $t = 1$, we obtain (8.8) for all Jensen measures σ, and for all $u \in \mathrm{Re}(A)$ such that $\int |{*u}|^p d\sigma = 1$. Since (8.8) is appropriately homogeneous in u, the condition $\int |{*u}|^p d\sigma = 1$ can be dropped, and the Kolmogoroff estimate is proved.

We have already given a proof of the M.Riesz estimate (8.9) in the case of the disc algebra, where duality allows us to deduce the estimate in the case $2 \le p < \infty$ from that in the case $1 < p \le 2$. Theorem 8.3 then shows that the estimate extends to Jensen measures.

The weak-type estimate (8.10) is proved in the case of the disc algebra by applying Theorem 7.12 to the positive and negative parts of u. For the general case, we express (8.10) in the form $\int H_c(u,{*u})d\sigma \ge 0$, where

$$H_c(x,y) = \begin{cases} |x| , & |y| < \lambda , \\ |x| - c/\lambda , & |y| \ge \lambda . \end{cases} \tag{8.13}$$

We have already remarked that the statements (ii) and (iii) of Theorem 8.3 are equivalent when H is only semi-continuous. Consequently (8.10) is valid also for all Jensen measures. □

Sharp Constants

A striking feature of the proof of Theorem 8.3 is that it points towards the extremal functions giving the sharp constants in the classical estimates. The proof shows that if an integral estimate fails, then it already fails for very special functions $u+i^*u$, namely, functions that are covering maps of Δ onto certain domains in the complex plane. Symmetry considerations may limit substantially the possible domains.

For instance, the radial behaviour of the function associated with the M.Riesz inequality convinces one that the domains to be checked are wedges symmetric about the x-axis. We are led then to consider the functions

$$g_\alpha(z) = \left(\frac{1+z}{1-z}\right)^\alpha, \quad z \in \Delta,$$

defined for $0 < \alpha < 1$. The function g_α maps the open unit disc Δ onto the wedge $\{\frac{-\alpha\pi}{2} < \arg w < \frac{\alpha\pi}{2}\}$, and $g_\alpha(0) = 1$. On $\partial\Delta$, the values of g_α lie on the rays $\{\arg w = \pm\alpha\pi/2\}$. In terms of the decomposition

$$g_\alpha = u_\alpha + iv_\alpha$$

of g_α into real and imaginary parts, this becomes

$$u_\alpha = \cos(\tfrac{\alpha\pi}{2})\,|g_\alpha|,$$

$$v_\alpha = \pm\sin(\tfrac{\alpha\pi}{2})\,|g_\alpha|$$

on $\partial\Delta$. Note also that $\int u_\alpha \frac{d\theta}{2\pi} = u_\alpha(0) = 1$, and $u_\alpha \geq 0$. If $\alpha < 1/p$, then $g_\alpha \in L^p(d\theta)$.

Let $1 < p < 2$, and let $\alpha < 1/p$. Since

$$|v_\alpha| = \tan(\tfrac{\alpha\pi}{2})\, u_\alpha,$$

we obtain

$$\left(\int |v_\alpha|^p \frac{d\theta}{2\pi}\right)^{1/p} = \tan(\tfrac{\alpha\pi}{2}) \left(\int |u_\alpha|^p \frac{d\theta}{2\pi}\right)^{1/p}.$$

Letting α increase to $1/p$, we see that the best constant c_p for the estimate (8.9) satisfies

$$c_p \geq \tan(\frac{\pi}{2p}) , \quad 1 < p < 2 .$$

The argument we have hinted at shows that in fact equality holds here. It is easier to obtain the best constant, though, by choosing the constants α and γ of (8.1) judiciously, and in the process a slightly better estimate will emerge.

We wish to choose $\gamma > 0$ as small as possible, so that there exists $\alpha > 0$ satisfying (8.1). For this, fix $\alpha > 0$, and consider the function

$$F(\theta) = [\sin^p\theta + \alpha\cos(p\theta)]/\cos^p\theta , \quad 0 \leq \theta \leq \pi/2 .$$

One computes that

$$F'(\theta) = \frac{p\sin^{p-1}\theta}{\cos^{p+1}\theta}[1 - \alpha g(\theta)] ,$$

where

$$g(\theta) = \frac{\sin((p-1)\theta)}{\sin^{p-1}\theta} .$$

Now

$$g'(\theta) = \frac{p-1}{\sin^p\theta}\sin((2-p)\theta) > 0 .$$

Consequently g is strictly increasing, and F' has at most one zero. Furthermore, the zero θ_0 of F' satisfies $1 - \alpha g(\theta_0) = 0$, or

$$\frac{\sin((p-1)\theta_0)}{\sin^{p-1}\theta_0} = \alpha . \tag{8.14}$$

Since F' changes signs at θ_0, θ_0 is a maximum value for F. We conclude that if some $\theta_0 \in [0,\pi/2]$ satisfies (8.14), then F attains its maximum at θ_0. In other words, if $\theta_0 \in [0,\pi/2]$ is fixed, and if α is defined by (8.11), then F attains its maximum at θ_0, i.e.

$$\frac{\sin^p\theta + \alpha\cos(p\theta)}{\cos^p\theta} \le \frac{\sin^p\theta_0 + \alpha\cos(p\theta_0)}{\cos^p\theta_0}, \quad 0 \le \theta \le \frac{\pi}{2}. \quad (8.15)$$

Now set $\theta_0 = \pi/(2p)$, and define α by (8.14). The estimate (8.15) becomes

$$\sin^p\theta + \frac{\sin^{p-1}(\frac{\pi}{2p})}{\cos(\frac{\pi}{2p})}\cos(p\theta) \le \tan^p(\frac{\pi}{2p})\cos^p\theta, \quad 0 \le \theta \le \pi/2 .$$

This leads to the following form of (8.9):

$$\int |{*u}|^p \frac{d\theta}{2\pi} + \frac{\sin^{p-1}(\frac{\pi}{2p})}{\cos(\frac{\pi}{2p})}|u(0)|^p \le \tan^p(\frac{\pi}{2p}) \int |u|^p \frac{d\theta}{2\pi},$$

$$1 < p < 2 . \quad (8.16)$$

Note that the estimate (8.16) tends to (7.4) as p increases to 2.

In any event, the estimate (8.16) shows that the best constants for the M.Riesz estimate (8.9) are

$$c_p = \tan(\frac{\pi}{2p}), \quad 1 \le p \le 2 .$$

By duality, we obtain as best constants

$$c_p = \tan[\frac{\pi}{2}(1 - \frac{1}{p})], \quad 2 \le p < \infty .$$

The functions g_α can also be used to show that the Kolmogoroff estimate of Theorem 7.10 is sharp, and that the

Zygmund estimate (8.7) is sharp, in that it fails for $\beta = 2/\pi$.

Now we turn to the weak-type estimate (8.10). For this, we study first an auxiliary function F .

Denote the strip $\{|\text{Im}(z)| < \lambda\}$ by S . Define $F(x) = |x|$ if $z = x + iy \notin S$, and extend F to S to be the Poisson integral of the function $|x|$ on ∂S . Then F is continuous and subharmonic, and F is symmetric with respect to both coordinate axes.

From the symmetry of the Poisson kernel for S , one sees easily that

$$F(x+iy) \geq F(iy) , \quad x + iy \in S . \tag{8.17}$$

It follows that $\frac{\partial^2 F}{\partial x^2}(iy) \geq 0$ for $-\lambda < y < \lambda$. Since $\Delta F = 0$ on S , we obtain $\frac{\partial^2 F}{\partial y^2}(iy) \leq 0$ for $-\lambda < y < \lambda$, and F is a concave function on the vertical interval $-\lambda < y < \lambda$. Since F is symmetric, while $F(i\lambda) = F(-i\lambda) = 0$, F must attain its maximum value over the interval at $y = 0$.

Now consider the function $F - |x|$ on S . It is superharmonic, it is harmonic except on the imaginary axis, it vanishes on ∂S , and it vanishes as $x \to \pm\infty$. Hence it attains its maximum over S on the imaginary axis. Since it coincides there with F , its maximum is attained at 0 . Hence

$$F(z) - |x| \leq F(0) , \quad z = x + iy \in S . \tag{8.18}$$

Now consider the function

$$H(z) = \begin{cases} |x| , & |y| < \lambda , \\ |x| - F(0) , & |y| \geq \lambda . \end{cases}$$

Then $F - F(0)$ coincides with H for $|y| \geq \lambda$, and the estimate (8.18) shows that $F - F(0) \le H$ when $|y| < \lambda$. By (8.17), $F - F(0) \ge 0$ on $\mathbb{R}$. Applying Theorem 8.3 to $h = F - F(0)$, we obtain $\int H(u,{*u})d\sigma \ge 0$ for all Jensen measures σ and all $u \in \mathrm{Re}(A)$. This yields the weak-type estimate (8.10) in the form

$$\sigma(\{|{*u}| \ge \lambda\}) \le \frac{1}{F(0)} \int |u|d\sigma \ , \quad u \in \mathrm{Re}(A) \ , \tag{8.19}$$

where σ is a Jensen measure.

We claim that the constant $1/F(0)$ appearing here is sharp. Indeed, suppose that the weak-type estimate (8.10) is valid with constant c/λ. Define H_c by (8.13), and let h_c be the upper envelope of the subharmonic functions on $\mathbb{C}$ that are dominated by H_c. By Theorem 8.3, $h_c \ge 0$ on $\mathbb{R}$.

Since $|x| - \lambda/c$ is a subharmonic function dominated by H_c, $h_c \ge |x| - \lambda/c$, and h_c must coincide with $|x| - \lambda/c$ for $|y| \ge \lambda$. By the maximum principle, $h_c \le F - \lambda/c$ on S. In particular, $0 \le h_c(0) \le F(0) - \lambda/c$. Hence $c/\lambda \ge 1/F(0)$, and the estimate (8.19) is sharp.

It remains to evaluate $1/F(0)$. For this, we consider the conformal map $(2\lambda/\pi) \log \zeta$ of the right half-plane $\{\mathrm{Re}(\zeta) > 0\}$ onto the strip S. Making the corresponding change of variable in the integral, we obtain

$$F(0) = \int_{\partial S} |x| d\mu_o = \int_{-\infty}^{\infty} \frac{2\lambda}{\pi} \left|\log|t|\right| \frac{dt}{\pi(1+t^2)}$$

$$= \frac{8\lambda}{\pi^2} \int_1^{\infty} \frac{\log t}{1+t^2} dt = \frac{8\lambda\kappa}{\pi^2} ,$$

where

$$\kappa = \int_1^{\infty} \frac{\log t}{1+t^2} dt = 1 - \frac{1}{3^2} + \frac{1}{5^2} - \frac{1}{7^2} + \dots$$

is the mysterious Catalan constant. The best constant in (8.10) is thus given by

$$\frac{c}{\lambda} = \frac{1}{F(0)} = \frac{\pi^2}{8\kappa\lambda} .$$

One can also conclude that (8.19) is sharp by verifying that (8.19) becomes equality when

$$f = u + i{*}u = \frac{2\lambda}{\pi} \log\left(\frac{1+w}{1-w}\right)$$

is the conformal map of Δ onto S .

Arens-Singer Measures

Cole's theorems have an analogue for Arens-Singer measures.

8.5 Theorem. Let D be a domain in the complex plane, let $z_o \in D$, and let H be a continuous, real-valued function on D . Then the following are equivalent.

(i) There exists a harmonic function h on D such that $h \le H$, and $h(z_o) \ge 0$.

(ii) If A is a uniform algebra, $\phi \in M_A$, and σ is an Arens-Singer measure for ϕ , then

$$\int (H \circ f) d\sigma \ge 0$$

for all $f \in A$ satisfying $f(M_A) \subseteq D$ and $f(\phi) = z_o$.

The proof runs along the same lines as the proofs of Theorems 7.2 and 8.3.

Note that if the domain D of Theorem 8.5 is simply connected, then h is the real part of an analytic function on D . Comparing Theorems 7.3 and 8.5, we see that in this case, an integral estimate is valid for all representing measures as soon as it is valid for all Arens-Singer measures.

Notes on Sources

As mentioned earlier, the extension of the classical inequalities to Jensen measures is due to B.Cole, and has been presented by him in various talks dating back to 1970. The proof of the M.Riesz Theorem, given at the beginning of this chapter, together with a determination of the best constant, is due to S.K.Pichorides[5]. Cole had independently discovered the best constant for the M.Riesz estimate.

The best constant for the weak type estimate was discovered by B.Davis[2]. Davis' proof, which depends on Brownian motion, is discussed by D.L.Burkholder in [1]. The proof we have given is close to the one given in the classical case by Albert Baernstein, II. It was modified to cover Jensen measures by K.Yabuta[6], who had independently discovered Lemma 8.1 as a vehicle for extending classical estimates to Jensen measures (cf. reference [12] of Chapter 7).

Davis[3] has also obtained the best constants for the Kolmogoroff estimate (8.8). Pichorides [5] has discovered, for each $\beta > 2/\pi$, the best constant γ that serves in the Zygmund estimate (8.7). The best constant for the Zygmund estimate for $\int \exp(\alpha|{*u}|)d\sigma$ given in Theorem 7.9 turns out to be

$$\frac{2}{\cos \alpha} - \frac{4}{\pi}\int_0^1 \frac{t^{(2\alpha/\pi)}}{1+t^2}\,dt = \frac{4}{\pi}\int_1^\infty \frac{t^{(2\alpha/\pi)}}{1+t^2}\,dt .$$

References

1. Burkholder, D.L. Harmonic analysis and probability, in Studies in Harmonic Analysis, J.M.Ash (ed.), *MAA Studies in Mathematics*, Vol.13, pp.136-149.
2. Davis, B. On the weak type (1,1) inequality for conjugate functions, *Proc. A.M.S.* 44 (1974), 307-311.
3. Davis, B. On Kolmogorov's inequalities $||\tilde{f}||_p \le C_p||f||_1$, $0 < p < 1$, *Trans. A.M.S.* 222 (1976), 179-192.

4. Gamelin, T.W. The polynomial hulls of certain subsets of $\mathbb{C}^2$, *Pac. J. Math.* 61 (1975), 129-142.
5. Pichorides, S.K. On the best values of the constants in the theorems of M.Riesz, Zygmund and Kolmogorov, *Studia Math.* 44 (1972), 165-179.
6. Yabuta, K. Kolmogorov's inequalities in the abstract Hardy space theory, *Arch. Math.* 30 (1978), 418-421.

9 · Moduli of functions in $H^2(\sigma)$

Let X be a compact Hausdorff space, let A be a uniform algebra on X, and let $\phi \in M_A$. We assume that ϕ has a unique representing measure σ on X. Let A_ϕ denote the kernel of ϕ, and let $H^p(\sigma)$ and $H^p_0(\sigma)$ denote respectively the closures of A and A_ϕ in $L^p(\sigma)$. In the case $p = \infty$, the closures are taken in the weak-star topology of $L^\infty(\sigma)$. The following facts about the function theory for $H^p(\sigma)$ are known and will be used in one form or another.

Every $f \in H^\infty(\sigma)$ can be approximated pointwise almost everywhere by a sequence $\{f_n\}$ in A satisfying $\|f_n\|_X \le \|f\|_\infty$. (9.1)

$$L^2(\sigma) = H^2(\sigma) \oplus \overline{H^2_0(\sigma)} . \tag{9.2}$$

$$\log|f(\phi)| \le \int \log|f|d\sigma , \quad f \in H^1(\sigma) . \tag{9.3}$$

The conjugation operator $u \to *u$ extends to a continuous operator from $L^1_R(\sigma)$ to $L^p_R(\sigma)$, $0 < p < 1$. (9.4)

If $u \in L^\infty_R(\sigma)$, then $\exp(u+i*u) \in H^\infty(\sigma)$. (9.5)

If $u \in L^1_R(\sigma)$ and $e^u \in L^2(\sigma)$, then $\exp(u+i*u) \in H^2(\sigma)$. (9.6)

The estimate (9.3) is the Jensen-Hartogs inequality, which is valid since σ is necessarily a Jensen measure. The property (9.4) follows from the Kolmogoroff inequality, once it is shown that $Re(A)$ is dense in $L^1_R(\sigma)$. Both (9.5) and (9.6) follow easily from (9.4). For detailed proofs, see [1] or [2].

We wish to address ourselves to the problem of characterizing the moduli of the functions in $H^2(\sigma)$. The property (9.6) provides a partial solution to the problem: if $w \in L^2_R(\sigma)$, then a sufficient condition that there exist $f \in H^2(\sigma)$ such that $|f| = w$ is that $\log w$ be summable. As a start, we might ask when the summability of $\log w$ is also a necessary condition in order that $w = |f|$ for some $f \in H^2(\sigma)$.

The summability of $\log w$ is a necessary condition, when A is the disc algebra, and $\sigma = d\theta/2\pi$. In this case, Szegö's Theorem asserts that $\log |f|$ is summable whenever $f \in H^2(d\theta)$ is not identically zero. Szegö's Theorem is proved by merely applying the Jensen-Hartogs inequality to the function f/z^m , where m is the order of vanishing of f at the origin.

Consider next the transplants $H^p((1+t^2)^{-1}dt)$ of the spaces $H^p(d\theta)$ to the upper half-plane. The algebra $H^\infty((1+t^2)^{-1}dt)$ can be identified with the algebra of bounded analytic functions on the upper half-plane, while the Poisson kernel $[\pi(1+t^2)]^{-1}dt$ is a representing measure for the point i . The space $H^2((1+t^2)^{-1}dt)$ is the closure of $H^\infty((1+t^2)^{-1}dt)$ in $L^2((1+t^2)^{-1}dt)$, and all the functions in $H^2((1+t^2)^{-1}dt)$ extend to be analytic on the upper half-plane. In this case, Szegö's Theorem asserts that $\log |f|$ belongs to $L^1((1+t^2)^{-1}dt)$ whenever $f \in H^2((1+t^2)^{-1}dt)$ is not identically zero.

A more complicated situation is provided by the algebras of generalized analytic functions associated with compact groups with archimedean-ordered duals. Let Γ be a dense subgroup of $\mathbb{R}$, and outfit Γ with the discrete topology. Let $G = \hat{\Gamma}$, the dual group of Γ , and let σ be the Haar measure on G . Each $a \in \Gamma$ determines a character χ_a on G , and we define A to be the uniform algebra on G generated by the characters χ_a , for $a \geq 0$, $a \in \Gamma$. Then $\phi(f) = \int f d\sigma$ defines a multiplicative linear functional on A , and in fact A_ϕ is the closed linear span of the characters χ_a , for $a > 0$.

We can embed $\mathbb{R}$ into G as a dense subgroup via the map $t \to e_t$, where e_t is the character on Γ defined by

$$\langle e_t, a \rangle = e^{ita} , \quad a \in \Gamma .$$

If $a \geq 0$, then $\chi_a(e_t) = e^{ita}$, $t \in \mathbb{R}$, extends to be bounded and analytic in the upper half-plane. Thus the functions in A are analytic almost-periodic functions in the upper half-plane. Similarly, for $f \in A$, $f_x(t) = f(x+e_t)$ is analytic almost-periodic in the upper half-plane above the coset $x + \mathbb{R}$, for all x in G . The functions in $H^2(\sigma)$ are analytic in the sense that $f_x(t) = f(x+e_t)$ belongs to $H^2((1+t^2)^{-1}dt)$ for almost all $x \in G$.

Since σ is a unique representing measure on G for ϕ , the facts (9.1) through (9.6) are valid for σ . However Szegö's Theorem fails. H.Helson and D.Lowdenslager[4] have shown there are nontrivial functions f in $H^2(\sigma)$ such that $\log |f|$ is not integrable. The characterization of the moduli of functions in $H^2(\sigma)$ remained an outstanding problem until 1973, when Helson[3] succeeded in establishing the following.

Helson's Theorem. Let Γ, G, σ and A be as above, and let $w \in L^2(\sigma)$ satisfy $w > 0$. Then the following are equivalent:

$$w = |f| \text{ for some } f \text{ in } H^2(\sigma) ; \tag{9.7}$$

$\log w(x+e_t)$ belongs to $L^1((1+t^2)^{-1}dt)$ (9.8)

for almost all $x \in G$;

wA is not dense in $L^2(\sigma)$. (9.9)

Actually, Helson and Lowdenslager had shown earlier in [4] that (9.8) and (9.9) are equivalent, while (9.8) follows from (9.7) by applying Szegö's Theorem to the functions $f_x \in H^2((1+t^2)^{-1}dt)$. Helson's contribution in [3] is to show that (9.8) and (9.9) imply (9.7).

In a seminar talk in 1974, B.Cole presented a simplified proof of Helson's Theorem. It turns out that Cole's proof extends to a more general context. The theorem we aim to establish is the following.

9.1 Theorem. Let A be a uniform algebra on X , and suppose that $\phi \in M_A$ has a unique representing measure σ on X . Suppose furthermore that no function in $H^2(\sigma)$ can vanish on a set of positive measure, unless it is identically zero. Let $w \in L^2(\sigma)$ satisfy $w > 0$. Then there exists $f \in H^2(\sigma)$ such that $|f| = w$ if and only if wA is not dense in $L^2(\sigma)$.

In other words, the conclusion of the theorem is that (9.7) and (9.9) are equivalent. Since the hypothesis on the zero sets of functions in $H^2(\sigma)$ is met in the case treated by Helson, Theorem 9.1 combines with the earlier work of Helson and Lowdenslager to yield Helson's Theorem.

The remainder of this lecture is devoted to proving Theorem 9.1. It will be obtained as a corollary of the more general Theorem 9.3. The idea of the proof is contained in the following Theorem 9.2. The Jensen-Hartogs inequality is crucial to the proof.

9.2 Theorem. Let A be a uniform algebra on X , and

suppose that $\phi \in M_A$ has a unique representing measure σ on X. Let M be a closed subspace of $L^2(\sigma)$ such that $AM \subseteq M$, and let E be a set of minimal σ-measure such that every member of M vanishes off E. Then there exists F in M such that $|F| = 1$ almost everywhere on E.

Proof. The proof breaks into four steps.

Step I. If the complex numbers $a_1, a_2, \ldots$ satisfy $\Sigma |a_j| < \infty$, then

$$\max_{1 \le j < \infty} |a_j| \le \int_{T^\infty} \log \left| \sum_{j=1}^{\infty} a_j e^{i\theta_j} \right| d\tau(\theta_1, \theta_2, \ldots) . \quad (9.10)$$

Here τ is the normalized Haar measure on the infinite torus T^∞.

Applying the Jensen-Hartogs inequality to $\bar{a} + \bar{b}e^{i\theta}$, we obtain

$$\log|a| \le \int \log|\bar{a} + \bar{b}e^{i\theta}| \frac{d\theta}{2\pi} = \int \log|ae^{i\theta} + b| \frac{d\theta}{2\pi} .$$

(In fact, one can establish easily the identity

$$\max(\log|a|, \log|b|) = \int \log|ae^{i\theta} + b| \frac{d\theta}{2\pi} ,$$

which already appears in an equivalent form in Jensen's paper of 1898 cited in Chapter 2.) Hence

$$\log|a_1| \le \int \log|a_1 e^{i\theta_1} + a_2 e^{i\theta_2} + \ldots| \frac{d\theta_1}{2\pi}$$

for fixed $\theta_2, \theta_3, \ldots$. By the uniqueness of Haar measure, $\tau(\theta_1, \theta_2, \ldots) = \frac{d\theta_1}{2\pi} \tau(\theta_2, \theta_3, \ldots)$. Integrating our inequality with respect to $\tau(\theta_2, \theta_3, \ldots)$, we obtain

$$\log|a_1| \le \int \log|a_1 e^{i\theta_1} + a_2 e^{i\theta_2} + \dots| \, d\tau(\theta_1,\theta_2,\dots) .$$

The same estimate applies to the other a_j's .

Step II. There exists h in M such that $h \ne 0$ almost everywhere on E .

Indeed, for $f \in M$, let $E(f)$ be the set on which f does not vanish, and set

$$\alpha = \sup\{\sigma(E(f)) : f \in M\} .$$

Choose a sequence $\{f_j\}$ in M such that $\sigma(E(f_j)) \to \alpha$. Let $g \in M$, and choose $\lambda_j \ne 0$ so that g/f_j does not assume the value λ_j on a set of positive measure. Then $E(g+\lambda_j f_j)$ coincides almost everywhere with $E(g) \cup E(f_j)$, so that $\alpha \ge \sigma(E(g+\lambda_j f_j)) = \sigma(E(g) \cup E(f_j)) = \sigma(E(f_j)) + \sigma(E(g)\backslash E(f_j))$. Letting j tend to ∞ , we find that $\sigma(E(g)\backslash E(f_j)) \to 0$, so that almost all points of $E(g)$ are included in $\cup E(f_j)$. Since $g \in M$ is arbitrary, we obtain

$$E = \bigcup_{j=1}^{\infty} E(f_j) . \tag{9.11}$$

Multiplying the f_j's by constants, we can assume that $\Sigma \int |f_j| d\sigma < \infty$. Then $\Sigma |f_j(x)| < \infty$ almost everywhere. For $\varepsilon > 0$, define

$$B_\varepsilon = \{x : \max_{1 \le j < \infty} |f_j(x)| \ge \varepsilon\} .$$

From (9.10) we obtain

$$\begin{aligned} \sigma(B_\varepsilon) \log \varepsilon &\le \int_{B_\varepsilon} \max_{1\le j\le\infty} \log|f_j(x)| d\sigma(x) \\ &\le \int_{B_\varepsilon} \int_{T^\infty} \log\left| \sum_{j=1}^{\infty} f_j(x)\, e^{i\theta_j} \right| d\tau(\theta) d\sigma(x) . \end{aligned}$$

In particular, for τ-almost all choices of $\theta_1, \theta_2, \ldots$, the function $\Sigma e^{i\theta_j} f_j(x)$ does not vanish on B_ε . From (9.11) we see that E is the union of the E_ε's . Hence $h(x) = \Sigma e^{i\theta_j} f_j(x)$ does not vanish on E , for almost all choices of $\theta_1, \theta_2, \ldots$.

Step III. There exists an f in M such that $\log |f|$ is integrable on E .

This step is the core of the proof. The idea is due to Cole, and a similar idea had been used by Helson.

Fix h as in Step II. Choose a sequence $\{\varepsilon_j\}_{j=1}^{\infty}$ of positive numbers such that $\varepsilon_1 = 1$ and $\Sigma \varepsilon_j < \infty$. According to (9.5), there exist $g_j \in H^\infty(\sigma)$ such that $|g_j| = 1/|h|$ whenever $1/(j+1) < |h| \leq 1/j$, while $|g_j| = \varepsilon_j$ otherwise. From (9.1) and the hypothesis $AM \subseteq M$, we see that $H^\infty(\sigma)M \subseteq M$. Hence $f_j = g_j h$ belongs to M . Moreover,

$$|f_j(x)| = \begin{cases} 1 , & \frac{1}{j+1} < |h(x)| \leq \frac{1}{j} , \\ \varepsilon_j |h(x)| , & \text{otherwise}, \end{cases}$$

and $|f_1(x)| = |h(x)|$ whenever $|h(x)| \geq 1$. Thus

$$\max_{1 \leq j < \infty} |f_j(x)| \geq 1 , \quad \text{almost all } x \in E .$$

Furthermore, $\Sigma |f_j(x)|$ converges to a summable function. Using Step I, we obtain

$$\begin{aligned} 0 &\leq \int_E \max_{1 \leq j < \infty} \log |f_j(x)| \, d\sigma(x) \\ &\leq \int_E \int_{T^\infty} \log |\textstyle\sum e^{i\theta_j} f_j(x)| \, d\tau(\theta_1, \theta_2, \ldots) d\sigma(x) . \end{aligned}$$

Interchanging the orders of integration, we find that the

summable function

$$f = \sum_{j=1}^{\infty} e^{i\theta_j} f_j$$

satisfies

$$\log|f| \in L^1(\sigma|_E)$$

for almost all choices of $\theta_1, \theta_2, \ldots$. Since $\Sigma \int |f_j|^2 d\sigma$ is finite, the series defining f converges in $L^2(\sigma)$, and f belongs to M .

Step IV. Choose f as in Step III, define $u \in L^1_R(\sigma)$ so that $u = -\log|f|$ on E , while $u = 0$ off E . Set $F = f \exp(u + i{*}u)$. Since $|F| = 1$ on E , it suffices to show that $F \in M$. For this, define $u_n \in L^\infty_R(\sigma)$ by

$$u_n(x) = \begin{cases} -\log|f(x)| , & -n < \log|f(x)| < n \\ 0 , & \text{otherwise} , \end{cases}$$

and set $F_n = f \exp(u_n + i{*}u_n) \in M$. By (9.4), ${*}u_n$ converges to ${*}u$ in measure, so that F_n converges to F in measure. Since $|F_n| \le \max(1, |f|)$, the convergence is dominated, and F_n converges to F in $L^2(\sigma)$.

□

For the next theorem, it will be convenient to denote the closure of a linear subset B of $L^2(\sigma)$ by [B] .

9.3 Theorem. Let A be a uniform algebra on X , and suppose that $\phi \in M_A \backslash X$ has a unique representing measure on X . Let $w \in L^2(\sigma)$ satisfy $w \ge 0$. Let E be a set of minimal measure such that [wA] includes all functions in

$L^2(\sigma)$ vanishing on E. Then

(i) there exists an f in $H^2(\sigma)$ such that $|f| = w$ on E and $f = 0$ off E;

(ii) there exists an f in $H^2(\sigma)$ such that $|f| = w$ if and only if E has full measure.

Proof. Let $[w\bar{A}]^\perp$ be the orthogonal complement in $L^2(\sigma)$ of the subspace spanned by $w\bar{h}$, for h in A. Then $g \in [w\bar{A}]^\perp$ if and only if $\int gwh d\sigma = 0$ for all $h \in A$. Evidently $[w\bar{A}]^\perp$ is a closed invariant subspace of $L^2(\sigma)$. Our hypothesis shows that all functions in $[w\bar{A}]^\perp$ vanish off E, and that E has minimal measure among sets with this property. By Theorem 9.2, there is an F in $[w\bar{A}]^\perp$ such that $|F| = 1$ on E and $F = 0$ off E. Since $\int Fwh d\sigma = 0$ for all $h \in A$, Fw belongs to $H_0^2(\sigma)$, by (9.2). Setting $f = Fw$, we obtain (i), and we also obtain the "if" statement of (ii).

Suppose there exists an f in $H^2(\sigma)$ such that $|f| = w$. Define a bounded function F so that $F = 0$ wherever $f = 0$, while $F = w/f$ elsewhere. Then $w = Ff$, and $[wA] = F[fA] \subseteq FH^2(\sigma)$. Evidently F is unimodular off E, so that the definition of E shows that $H^2(\sigma)$ includes all functions in $L^2(\sigma)$ that vanish on E. However, the only real-valued functions in $H^2(\sigma)$ are the constants. (If $u \in H_0^2(\sigma)$ is real, then $u^2 \in H_0^1(\sigma)$, so that $\int u^2 d\sigma = 0$, and $u = 0$.) Since $\phi \notin X$, σ cannot be a point mass. We conclude that E has full measure. Cliffs, 1969. □

Proof of Theorem 9.1. If $\phi \in X$, then σ is the point mass at ϕ, and the theorem is obvious. We assume then that $\phi \in M_A \backslash X$. Pick E as in the statement of Theorem 9.3. We must show that E has full measure if and only if $[wA] \neq L^2(\sigma)$. From part (i) of Theorem 9.3 and the hypothesis on the zero sets of functions in $H^2(\sigma)$, we see that either E has full measure, or E has zero measure. By the definition of E, $[wA] \neq L^2(\sigma)$ in the former case, while $[wA] = L^2(\sigma)$ in the latter case.

□

References

1. Browder, A. *Introduction to Function Algebras*, W.A. Benjamin, Inc., New York, 1969.
2. Gamelin, T.W. *Uniform Algebras*, Prentice Hall, Englewood Cliffs, 1969.
3. Helson, H. Compact groups with ordered duals IV, *Bull. London Math. Soc.* 5 (1973), 67-69.
4. Helson, H. and Lowdenslager, D. Prediction theory and Fourier Series in several variables II, *Acta Math.* 106 (1961), 175-213.

List of notation

$\mathbb{R}$	real line
$\mathbb{C}$	complex plane
Δ	open unit disc in complex plane
Z	integers
Z_+	strictly positive integers
$\bar{E}$	closure of E
∂E	topological boundary of E
E^o	interior of E
X	compact Hausdorff space
$C(X)$	continuous complex-valued functions on X
$C_R(X)$	continuous real-valued functions on X
A	uniform algebra on X
M_A	maximal ideal space of A
∂_A	Shilov boundary of A (p.24)
$\mathrm{Re}(A)$	space of real parts of functions in A
A^{-1}	invertible functions in A
$\log\lvert A^{-1}\rvert$	$\{\log\lvert f\rvert : f \in A^{-1}\}$
A_ϕ	kernel of $\phi \in M_A$
$A \otimes P$	algebra of Hartogs polynomials (p.58)
A_E	closure in $C(E)$ of the restriction algebra $A\vert_E$ (p.65)
$\hat{E}$	A-convex hull of E
$H^\infty(D)$	bounded analytic functions on D (p.46)
$M(D)$	maximal ideal space of $H^\infty(D)$ (p.46)
$A(\Delta)$	disc algebra (p.30)
$R(K)$	closure in $C(K)$ of rational functions with poles off K (p.34)
$\mathcal{O}(K)$	functions analytic in a neighbourhood of K (p.88)

$H(K)$ closure of $\mathcal{O}(K)$ in $C(K)$ (p.88)

$\mathcal{H}(K)$ closure in $C_R(K)$ of functions harmonic in a neighbourhood of K (p.37)

$\log^+ u$ $\max(0, \log u)$

u^* upper semi-continuous regularization of u (p.65)

$^*u$ harmonic conjugate of a harmonic function u; abstract conjugate function of u (p.107)

$\tilde{u}$ solution of the $\mathcal{R}$-Dirichlet problem (p.9); solution of the A-Dirichlet problem (p.29, p.56)

$\check{u}$ solution of the generalized Dirichlet problem (p.78)

$\tilde{u}$ (p.102)

$H_u(z)$ complex Hessian matrix of u (p.83)

Δu Laplacian of u

δ_ϕ point mass at ϕ

$\mathrm{supp}(\sigma)$ closed support of σ

$\hat{\nu}$ Cauchy transform of ν (p.34)

V_ν negative of logarithmic potential of ν (p.35)

$H^p(\sigma)$ closure of A in $L^p(\sigma)$, $1 \le p < \infty$

$H^\infty(\sigma)$ weak-star closure of A in $L^\infty(\sigma)$

$H^p_o(\sigma)$ closure of A_ϕ in $L^p(\sigma)$ (weak-star, if $p = \infty$)

$[B]$ closure of B in $L^2(\sigma)$ (p.153)

$\mathcal{R}$ (p.1)

$\partial_{\mathcal{R}}$ Choquet boundary associated with $\mathcal{R}$ (p.10)

$\mathcal{U}$ (p.2)

$\mathcal{S}$ $\mathcal{R}$-envelope functions (p.5)

$\mathcal{S}_C$ continuous $\mathcal{R}$-envelope functions (p.8)

$\prec$ (p.16)

Index